MW00776642

MATERIAL AND MIND

CHRISTOPHER BARDT

The MIT Press
Cambridge, Massachusetts
London, England

This book was set in Adobe Garamond Pro and Berthold Akzidenz Grotesk by Jen Jackowitz. Printed and bound in the United States of America.

Library of Congress Cataloging-in-Publication Data

Names: Bardt, Christopher, author.
Title: Material and mind / Christopher Bardt.
Description: Cambridge, MA : The MIT Press, [2019] | Includes bibliographical
    references and index.
Identifiers: LCCN 2018054294 | ISBN 9780262042727 (hardcover : alk. paper)
Subjects: LCSH: Design--Psychological aspects. | Materials.
Classification: LCC NK1520 .B37 2019 | DDC 745.401/9--dc23 LC record available at
    https://lccn.loc.gov/2018054294

10 9 8 7 6 5 4 3 2 1

For the B

A good traveler has no fixed plans, and has no intent on arriving.

—LAO TZU

Contents

Acknowledgments

This book has had a long gestation, and many more people have contributed to my thinking and writing than I realized as it was developing. First I want to recognize my parents, who profoundly influenced me even as they carried on with their daily lives. My mother, always engrossed in a book, more often than not a biography of an artist, passed to me her love of the arts, books, and reading. My father, a scientist, a tinkerer by nature, seemingly repurposed everything: he might use a spent brass cartridge to repair broken eyeglasses, gently hammering the tube to magically fit the parts of the arm together. He taught me how to think by making. Such influences are osmotic: they seep in, quietly shaping us, and for that reason I am also grateful to friends and colleagues with whom I've had enduring relationships.

I wish to acknowledge those who participated directly in fostering this book. After years of gathering notes, which seemed to amount to gathered notes, not much more, I was encouraged by then-RISD President John Maeda to write a book. For that push I am grateful. The ideas of this book slowly grew from years of discussions about architecture, material, and the intersection of reasoning and making and the imagination. I thank my many colleagues from the various disciplines at Rhode Island School of Design for their input and insights over those years. I will single out designers Seth Stem and Robert O'Neal, painter Alfred De Credico, and architect Friedrich St.Florian as colleagues, friends, mentors, and deeply sensitive thinkers and makers.

I have also been fortunate to have had wonderful students, some of whose work appears in this book. I won't name them all to avoid leaving anyone out, but I do want to acknowledge their original and vital contributions to the topic of material and mind.

Our practice, 3six0 Architecture, has contributed many of the ideas of the book and demonstrated them in built form. I thank all our staff, especially Jack Ryan, who has been a steady leader, friend, and creative contributor, as well as Matt Osborn and Alice Berresheim.

My first material experiments were done while I was in graduate school studying under John Hejduk, and I am grateful for the freedom he gave me to experiment and fail. Other professors at that time—Robin Evans, Donald Flemming, Paul Rotterdam, and Stephen Jay Gould—have, to this day, had an outsized impact on the thinking that arose from writing this book. The many long discussions and exchanges with my friend, fellow architect, thinker, and maker Peter Lynch served to underpin much of what this book is about. Friend and poet Stuart Blazer taught me that words are like material; they, too, can resist. Stuart, along with John Duke, Joan Richards, and John Connell, were patient and thoughtful readers of the full manuscript, and I am thankful for their keen insights and suggestions.

A RISD professional development fund grant and additional funding from the RISD Department of Architecture and the Division of Architecture and Design are gratefully acknowledged. Department head Laura Briggs gave early and steadfast encouragement throughout the writing and editing process. Department colleagues Hansy Better and Petra Kempf helped bring the manuscript to the publishing world, and students Chloe Renee Jensen, Margaret Kiladjian, Julie Kress, and Adelaide MacKintosh assisted with drawings, citations, and other prepublication work.

I wish to thank the many generous respondents to my requests for images, especially Barbora Benčiková, Daniela Hammer-Tugendhat, Alexandra Timpau, Heli Ojamaa, Denise Schmandt-Besserat, and Linda Pollak.

Executive Editor Robert Prior of the MIT Press was an early and gracious supporter of the manuscript, and has my gratitude for bringing this work to fruition. My editor, Scott Cooper, worked closely with me for the past three years, and I am deeply indebted to him for his guidance, patience, and wonderful ability to bring forth the best of his authors (at least this one). Scott believed in the book, even as a raw manuscript, saw in it something others didn't, and never wavered in his efforts to bring it to press. I am very fortunate to have worked with Scott.

Lastly, my greatest indebtedness is to Kyna Leski, exemplar of a beautiful intelligence, dear friend, life partner, partner in practice, colleague, and wife. As Kyna was writing her remarkable book *The Storm of Creativity*, the notion of my writing this book slowly emerged; her quiet osmotic influence has and continues to be the greatest of gifts.

Christopher Bardt
Providence, Rhode Island
September 26, 2018

INTRODUCTION

Bending spoons, "mind over matter," psychokinesis—this is the stuff of dreams and cinema and trickery. Since the nineteenth century, scientific researchers have investigated the possibility of acting on physical material solely with mental powers, but they have yet to find anything to support such an "action at a distance" phenomenon. The notion of using one's mental powers over matter persists, but the only credible arena in which the mind has control over matter is the corporeal. Sitting on a bed of nails and walking on hot coals are celebrated because they demonstrate the mind's ability to control and manipulate the material of its own body.

The fact that spoons can't be bent by the mind, though, doesn't mean material is impervious to the mind's volition. Since the dawn of technology, we humans have strived to reshape material, to "bend it" to our will. Our bodies, our hands, were the first agents of the mind; they have been replaced in the past few centuries by mechanical and now by digital devices. As we continue to transform the whole of our physical environment into a product of the mind's increasingly disembodied activities, it only increases our confidence in our mind's and thought's dominance over physical reality.

This transformation is not yet complete. As we enter the Anthropocene, the first geological age in which the chief agent of climactic and terrestrial change is human activity, our relationship to the material world can no longer afford to be one that assumes thought precedes and rules over physical material. Is it possible that the inverse is true as well?

The precedence of thought envelops us so completely that we hardly notice it is the foundation on which disciplines have been constructed. Archaeologists, for example, examine artifacts, "material culture," to trace back the rites, customs, and beliefs of past cultures—their state of mind— all the while assuming that every such artifact is a representation of mental activity. But what if the act of making those artifacts is what brought *thought* into existence? Do materials or our physical surroundings influence the mind?

It is sometimes supposed that thoughts at a bus stop will differ from thoughts while standing in a meadow, or that making your bed each morning might help structure your life, or that more daylight leads to better learning in classrooms. But the connections are vague. Underlying these suppositions is an assumption of a *passive* mind mysteriously influenced by its physical surroundings. Pronouncements about altering the physical context to affect one's mind teeter somewhere between science and self-help; the relationship between cause and effect eludes direct measure. As a result, the sciences, built on a foundation of causality, look in vain for specific relations between environment and mental states.

The French novelist Marcel Proust (1871–1922) captures the ambiguity of experience in a description from *In Search of Lost Time*:

> I learned for the first time that the variations in the importance which a pleasure or a pain has in our eyes may depend not merely on this alternation of two moods, but on the displacement of invisible beliefs, such, for example, as make death seem to us of no account because they bathe it in a glow of unreality, and thus enable us to attach importance to our attending an evening party, which would lose much of its charm for if, on the announcement that we were sentenced to die by the guillotine, the belief that had bathed the party in its warm glow was instantly shattered . . .[1]

The answer is less vague when we switch from physical *context* to physical *material*. Do our actions on the latter make us think and imagine differently? It is less vague precisely because in some fields the medium of work is literally the medium of thinking. Take the fabric away from the clothing designer or tailor, and their medium of thinking disappears as well. Vocations are in part defined by their means, which can be rephrased as the statement that *how you work determines what you think.* The unique *way* one works contributes to *how* one thinks: a carpenter differently than a lawyer; an artist differently than a writer; an engineer differently than a gymnast. Only when looking across many fields and forms of activity and the medium, tools, and materials of each individual activity does a picture emerge of a constant exchange between doing and thinking. And the greater the awareness of and respect for this exchange, the greater the potential to extend thinking beyond the habits of the mind that each vocation symbiotically forms with the means and materials it uses. This is the idea that underlies the frequent comment, "I don't know until I've tried." This is true for all doers in all disciplines, not only for the artists and designers who may have the greatest sense of this connection.

This implies that the concept of the mind as an encapsulated and autonomous consciousness needs revising. Philosophers/cognitive scientists Andy Clark and David J. Chalmers suggested in their groundbreaking 1998 essay "The Extended Mind"[2] that the mind is not encapsulated but extends outward, engaging and enacting itself with and in the environment. Their idea is at the center of the concept known as "extended cognition," which holds that the mind and its processes extend beyond the body into the environment within which an organism is embedded, and hence include aspects of the organism's interactions with that environment. Such a model of the mind corresponds with the notion of material informing thought, especially through action.

Throughout human history, the interactions of the mind and the material world have been mediated by the agencies specific to our species: language, images, and physical making. To see and understand how materials form, influence, and guide our thinking, we have to look through many disciplines, from anthropology to philosophy to the arts and design—in a sense, using each discipline as a lens to render partially visible (in the terms of that discipline) how a given material or medium interacts with and shapes our mind and its thoughts. Only by drawing from well beyond the specific world of the material-based arts can the relationship between material and mind be shown in the realms where they come together—in the forms of language, images, and making—and only such a broad, liberal (in the classical sense) approach can reveal how material and mind interact through surrogates such as metaphor, representation, projection, analogues, tools, and models.

Attempting to understand the relationship between material and the mind by studying only their correlate disciplines, material science for the former and psychology for the latter, is surely bound to disappoint given the lack of common ground. Nuclear physics offers a lesson in its pursuit of increasingly unobservable subatomic particles. Having long moved past observability as evidence for the existence of quarks, muons, and bosons, physicists now rely on indirect means such as symmetries, trajectories, energies, and other traces of behavior to "see" the invisible. Similarly, it is the *traffic* between material and mind that yields riches and evidence of their mutual influence. Surrogates such as metaphors act as go-betweens that allow the mind to extend itself and material to "enter" the mind. By looking across disciplines, such unselfconscious activity can be made more explicit. A simple example can be seen in the everyday device of metaphor, which often literally "bridges across" by ascribing material and physical attributes to otherwise abstract mental conditions, as in this example from American writer Charles Brockden Brown (1771–1810), who drew on the behavior of the physical material *water*: "My thoughts flowed with tumult and rapidity."[3]

We barely notice the flow between thinking and materials—that is, until something disrupts it. In scientific research, such "accidents" are associated with discovery. A Japanese engineer accidentally leaves a hot iron on his pen and, seeing the ink ejected, conceives of the ink jet printer. Harry Brearly, a metallurgist in Sheffield, England, trying to make stronger steel that can resist erosion in gun barrels from the friction of firing bullets, ends up accidentally creating stainless steel. In the arts and design, "accidents" are associated with creativity and the imagination, so much so that they are an integral part of the artistic process, which then raises the question of where exactly the imagination is located. If it is exclusively a mental operation, why then do materials—the "accident"—so often play a triggering role? That is a fundamental question this book explores: How are ideas, the imagination, and creativity influenced by and intertwined with physical material?

Materials are like a boulder in the road, resisting and disrupting suppositions, especially those being acted on; materials provoke, spark, and activate our imagination and creativity. A clue to imagination's external locus can be found in evolution: toolmaking, language, and visual art arguably codeveloped as a result of and resulting in increased visual memory (and imagination).

One of the oldest of human artifacts is a baboon bone, the so-called Ishango bone dating back about 20,000 years. It has a piece of quartz deliberately attached to its end; along its length, there are three columns of what some archaeologists have interpreted as tally marks. More than just a tallying, though, the marks are grouped and ordered, strongly suggestive to me of a mathematical mind, and thought by some scholars (although by no means all) to be an attempt to make sense of the moon's waxing and waning cycles. Regardless of the marks' precise purpose, the act of incising was a significant action, one that might be characterized as the mind extending itself into material or, conversely, enacting itself from the notches made in the bone, notches that needed the hardness and sharpness of the quartz and the give of the bone to be recorded. Given that concepts for numbers didn't exist at that time, the notches extend the mind's capacity for counting (beyond the innate three or four things) and for conceptualization, and in doing so the bone transforms into an instrument for thinking.

Design is a specific and revealing interaction of material and our mind that focuses on the ways in which humans make art, work with materials, and use tools. Design and the design imagination are influenced by the medium and materials used in the design process, which is certainly of great interest to designers. But design is also what we are surrounded by; the scrim of daily life consists almost solely of designed and manufactured materials and artifacts. Our interaction with this fabricated reality and how it affects our mind is consequential as we rapidly reach the limits of our physical resources. How we continue to shape our world as it continues to shape us is an important question for everyone. Just how this shaping is meaningful will lead us to tell new stories, ones that require imagination and empathy.

We humans see meaning in everything. Meaning articulates the ways we try to understand, anticipate, and render the places in the world we observe and make. Meaning resides in the stories we tell ourselves and we tell others (through a host of media). It is found in stories created to structure the rich, complex, and contradictory relationships between materials and our minds. Stories and their meaning bind concepts and material, transforming both.

Material is made meaningful by our transformation of it, but also by its capacity for memory. I'm not speaking of carved tablets and memorials; I am referring to material's inherent properties of holding time. For instance, a piece of granite is a result of its own making, a record of immensely geologic deep time—so deep that its link to prehistory was not conceived of until 200 years ago. But not all material memory stems from natural processes.

I remember walking in a light drizzle on a winter day in Warsaw, Poland, some decades ago, sensing something gloomy in the ubiquitous square cement pavers that covered the surface of every sidewalk. The color of the wet pavers was a gritty gray, darker than the cement I knew. Later, I found out that much of this smashed and leveled city of World War II infamy had been hauled off, pulverized, and reconstituted into these very pavers. In Warsaw, every footstep presses on an apparition of the tragic events of more than a half-century ago, bringing them back to our time. These ordinary, silent pavers transmit memory, events, displaced lives, and grief. There is no beauty in these pavers—far from it—but they make the city real and painfully complete.

Material provoking thought and memory is demonstrated, too, by an example from my institution, the Rhode Island School of Design (RISD). The building in which I teach, a nineteenth-century converted warehouse, retains its original wood floors, everything about which is wrong. Gouged, warped, stained, chipped, painted in spots, discolored, patched, they are the kind of floors where no one takes notice of the various daily abuses.

They have served as a handy surface to cut on; they sometimes backstop holes being drilled into some model part. No one seemed to care much until an alum who had become successful enough to be a major donor asked to acquire the old wooden floors to install in his own office, three thousand miles away, because he loved "that RISD feeling." We surprised ourselves, learning instantly how much these floors mean to us. We discovered that all those faults and defects are memories—and beloved ones at that. Needless to say, the floors remain unharvested, never to be exchanged for some characterless commercial replacement. The story illustrates how another material, also under foot, dwells in the mind.

These are two examples, Warsaw and RISD, of material evoking memory; of matter over (and under) mind.

* * *

Please indulge me this note to the reader.

To begin reading a book is to entrust oneself to a journey that many times has a known destination. That such a destination exists is a reasonable expectation. Some journeys, though, begin with destination unknown or uncertain.

Journeys may also involve some risk and surprises along the way. A sudden rain shower, a wrong turn in the road, a canceled reservation—these can be stressful annoyances or the beginning of adventures that transform and overturn expectations. It all depends on the traveler.

Reading this book should be one of those journeys. So, too, is my central theme: that working with material—every material—is a journey, more interesting, without a destination, full of surprises and stubbornness, and, most of all, deeply transformative of our state of being.

I begin with a simple premise: that our mind and thoughts are continuously formed and influenced by the physical material with which we work and by which we are surrounded. Reading this book is intended to be analogous to working a piece of material. Both book and material will resist a little, demand a degree of participation and forbearance, proceed in a nonlinear fashion, and—I hope—take the mind on an unanticipated and adventurous journey.

The book crosses a lot of disciplines; readers who are experts in one part of the discussion may find themselves novices in another. Although I've made every effort to avoid discipline-specific jargon when possible, the vocabulary here can be as demanding as crossing borders can be to travelers' language skills: despite its specificity, a word such as "resistance"—important to the book's theme—will mean one thing to an electrical engineer and another to a political scientist. The difficulties of crossing disciplines evoke the pejorative shadings of *dilettante*, one who knows less and less about more and more, nothing about everything. But crossing disciplines is necessary to see questions from multiple points of view, and in the case of material, to move between many points of view in order to understand how exactly the mind and material are entangled.

The familiar phrase "thinking outside the box" makes a metaphor of a need to think unconventionally, implying that familiar thoughts are contained (in a box) and unconventional and imaginative thoughts are found outside of containment. "Thinking outside the brain" might be another way to put it, which, as this journey of a book explores, is exactly what happens as we think and imagine and feel through our engagement with material. While not the destination of this book, it is its path, what's needed to begin.

1 THINK OR DO

The presumption that we first think of what we *will* do forms the basis of most disciplines and systems of education. The other way around—to have thought informed from or by action—is given little credence. Yet, as everyone who works with physical material knows, thoughts continuously emerge from action and physical work. These thoughts may be considered lesser, more like hunches than the thoughts of "pure" thinking, but they are thoughts nonetheless. Insights, understanding, intuition, grasping— they may all be classified as "soft thoughts" more closely associated with feelings and empathic sensibility, with "I get it" and "aha"—but does their "softness" make them lesser thoughts?

In human evolution, there's evidence to support a nonhierarchical relation between thought and action. As our species evolved, the expanding cerebral cortex (the thinking part) was accompanied by a growing cerebellum (the locus of motor control), suggesting that the newly freed hands were engaging in technical activities that were increasingly elaborate. One might argue that the development of the articulated hand drove the enlargement of the cerebral cortex, and that the brain should be viewed as a neurological continuum between gray matter and its extensions in the body.[1] Either way, there is ample physiological evidence of body-brain interdependence.

Beyond the physiological evidence, this also corresponds with intellectual discourse: the centuries-long philosophical debate over mind and matter, material and materiality, medium and message, form and content, has been propelled by the ambiguities surrounding the definition of and relationship between things and ideas. From Plato and Aristotle to Hegel and Kant, the debate has been ceaseless between the primacy of matter and the primacy of the mind or spirit. This continues today, running through other disciplines as well as philosophy. For instance, art making in the 1960s contended with a dramatic shift from considering material as secondary to form and idea to the new minimalism, which focused on materials and process as the locus of significance. While this book draws from these discussions, it is decidedly not intended to resolve philosophically the proper relation of, say, material and spirit. It aims instead to explore how action on and with material, through the lens of varied disciplines, can and has informed thinking and imagining, and ultimately has shaped the making and designing of our surroundings.

A glance at the etymology of *material* and *matter* gives us some clue. They share the same Latin root *materia*—meaning source, origin, mother, the substance from which something is made. But there is also a Doric Greek word from antiquity, *dmateria*, and there is an etymological theory involving that earlier word that figures here. It is the word that helps form the Proto-Indo-European root *dom-*, from which we get the Greek *domos* and the Latin *domus* ("house" or dwelling)—with other connections to words for the eventual English words "timber" and "wood."[2] By this theory, these two words create an interesting juxtaposition of an abstract and concrete basis that also invokes a temporal relationship. I think of it this way: whereas "materia" *precedes* any physical form, "dmateria" *anticipates* form. By the fourteenth century, roughly along these lines, the two terms had become differentiated: *matter* became associated with situation, circumstance, substance, content, cause, what something was made from; *material* carried with it an implication of potential to form, to be shaped in the future for useful ends (as in the raw *material* for, say, a novel).

Material is made of matter, but matter isn't made of material. Matter is homogeneous and uniform; material has physical traits and characteristics and limits.

The term *mind* originally referred to memory, but over time has come to include awareness and intention—the consciousness of the present and future, respectively, having been added to that of the past. *Mind* also finds definition as the faculty of imagination, that is, our *ability to form mental concepts that are independent of the senses.* This has led to an understanding of the imagination (and thought) as mental processes free of influence from the senses. This corresponds to our purest sense of the imagination as being unburdened, unlimited, and untouched by the world around us. Consistent with this implied autonomy, *mind* is also commonly framed in dualities such as mind/body, mind/matter, mind/spirit, all reflections of its nonmechanistic, immaterial, and self-aware nature.

Medium, for our purposes, refers to *means* or *agency*, and is not limited to material means but includes language, sound, the body, and images both physical and digital. It is much broader than its meaning in the art world as a reference to the specific materials used to create a work (e.g., "her painting medium is oils").

The term *medium* and its plural *media* imply a primarily transmissive function, a one-way street supporting the movement of intention to action to desired outcome, and rarely if ever is the medium conceived as a participating agent accompanying and influencing the process. A physical *medium* is even less likely to be considered for what it transmits *back* to us and for its effects on thought and imagination.

We almost never acknowledge that thought is formed from material. Education's overall emphasis on learning through language reflects this certitude. "Thinking" disciplines are exclusively language-based, segregated from image and making as sources: lawyers don't knead clay to develop an argument, and accountants don't draw pictures to balance the books. How working directly with material gives rise to thoughts is rarely, if ever, associated with the actual process of how we think and understand. Why is this?

Our common sense tells us that thinking is done with language. Reflection, contemplation, and planning are modes of thinking for which language provides structures. This is not the only mode our thoughts operate in—visual thinking is fundamental as well. But visual thinking is not considered "rational" or as credible a form of reasoning as language. Nevertheless, visual thinking and language-based thinking influence each other. How we think conditions how we see, and vice versa. And this reciprocal relationship isn't limited to vision; all our senses interact richly with thought, even though thought seems curiously unaware of this.

CONCEPTS AND MATERIAL

It wasn't always like this. In the beginning, before the widespread use of written language, concepts resided in material. A stone tool was not only a tool, but also the very idea of a tool—the idea and the stone of the tool, its material, were one and the same. Ideas had to be made, enacted, and painted into existence to form and perpetuate themselves. This explains, in part, the extraordinary level of craft and care that went into the making of cultural artifacts in the distant past.

When Frollo, the archdeacon in Victor Hugo's *Notre-Dame de Paris*,[3] declares, "This will kill that"—"this" being the book on his desk, the product of a printing press, and "that" being the cathedral—he is recognizing that buildings would no longer be the register of mankind, the edifice of ideas "read" and lived through by each civilization. Instead, the coming torrent of printed books would free ideas from material and, in doing so, become imperishable.

When the printed word pulled concepts away from material, words lost their original association with matter. Ideas were more freely formed with words, from thought alone. Now, ideas formed from words dominate and obscure those formed through working material, and ignore that concepts themselves are defined in great part by the traits of the medium in which they reside, whether in matter or words. The etymology of "concept" traces this dematerialization through history: the first sense of "concept" coming from the past participle stem of the Latin *concipere* meaning "to take within" (as in physiologically conceiving) and underscoring embodiment, which later became "to grasp," with the modern sense finally emerging in the fourteenth century, "to take into the mind."

The shift in authority from material to mind, from physical edifice to word, has gone so far as to put us at the threshold of digitally printed buildings, furniture, clothing, and tableware, promising to define every human artifact as the output of disembodied digitized thoughts. Consequently, material has lost its significance and meaning, and it is no longer considered the primary wellspring of creativity.

Even the fine arts and design, traditionally so material-bound that they organize their disciplines accordingly (glass, metals, ceramics, painting), are not immune to excluding physical material from the thinking process. Digital plotters, software, printers, CNC machinery, scripting, and 3D printers are ubiquitous in art schools and the studios and workshops of practicing artists and designers.

Is this a problem? In many fields, no. If fact, it is quite the opposite, with today's biggest advances in science and engineering being driven by computation. Take the internal-combustion, gasoline-fueled engine in automobiles, for example. Metallurgists, physicists, and mechanical engineers continue to refine the physical engine, but it is software that is bringing the greatest performance improvements. The engine's refinement from improvements to crankshafts, pistons, tubes, hoses, and bearings has slowed, but its overall evolution has been accelerated with controllers, sensors, lines of code, and automated adjustments taking the lead. Brawn is acquiring a brain.

It is easy to spot vocations that have never undergone a shift away from material or the medium in which they have always been conceptualized. Just imagine trying to write a computer script to dream up a new recipe for a chicken dinner. There seems to be no trend in the culinary arts to stop forming concepts from food itself—good news for those of us who love great food. The same can be said for other arts such as theater, dance, music, as well as crafts—for instance, while much glassblowing is already an automated process, a human glassblower relies on tacit knowledge to conceive and enact a blown work, the result of a unique dialogue between maker and material for each irreplicable piece. The dialogue is necessary because material, in this case glass, cannot be envisioned according to any fixed concept. The behavior of the molten glass—its changing viscosity under heat, weight, and pliability—is something that cannot be held in mind as idea. Even transparency, the concept most often ascribed to glass, is far too simplistic (as the mind holds it) to predict how a blown piece will catch, refract, transmit, and reflect light in all its rich complexities.

The glass being worked by the glassblower cannot be reduced to words or ideas—or "gotten" in the way a concept can be "gotten."

Phrases such as "I see now" and "I see what you mean" underscore the privileged relationship between the mind and the eye, and the assumption that understanding is analogous to seeing. The corollary to this is that incomprehension is akin to blindness, to feeling one's way in the dark— exactly what engaging with materials forces upon us. This may contribute to the uneasy relation between those who work with material and those who "see" through the distancing instruments of language and theory.

So why bother? Working with materials can be annoying. They slow us down—we can think more quickly than we can manipulate them, and they don't behave as we want them to. It's unlikely you will set out to carve some wood in the deliberate quest for an original insight. The inconvenience of it all is enough for us to be attracted to the promise of ease and control offered by software.

The assumption underlying this annoyance with manipulating material is that material's role is simply one of propagation, an agent of transmission, a medium for an intention, for thought, for fully formed desire or will. Were this true, bytes and words would be a far better thinking medium than materials. But materials do more than transmit. They derail assumptions. They resist propagation. They create doubt. They generate new ways of thinking, of imagining, of inventing.

THOUGHT AND ACTION, ACTION AND THOUGHT

Our actions on materials are a projection: we intend, we project, and we act. Materials react. What gets projected out through material? And what does material pass back to us?

The automotive industry offers some insights. Despite access to the most powerful design tools digital technologies have to offer, automotive design continues to rely on a hand-guided design sensibility. Before the digital age, auto manufacturers had a long tradition of designing cars by sculpting full-size clay models. It was widely assumed that the introduction of digital visualization software and digitally controlled milling machines would bring a quick end to hand-sculpted automobile modeling. It hasn't. Designers tried working exclusively with 3D software but found their results to be lifeless and difficult to evaluate.[4] Although they often couch it in language bordering on the sentimental, both designers and modelers speak of the "soul" and "transmission of emotion"[5] as key ingredients of car design—and say that only hand forming is capable of this almost magical exchange between author(s) and material.

Still, even with the persistence of clay modeling, software and computer-guided milling are also prominent in the modeling process; this is not an either-or situation. A hand sketch will often be digitized and altered, and a rough computer-milled representation of a digital model can serve as a starting point for a handmade clay model; but the final forming is all by hand or, more accurately, by the complex and immeasurable give and take between material and hand. The conceptualization of a surface is formed by a partnership between the volition of the sculptor and the material itself, its nature, and its character. For this reason alone, the auto industry refuses to approve production without first completing full-size handmade clay models.

In the early days of the automotive industry, engineering rationale and manufacturing techniques derived from wood coach-building traditions tended to limit designs. Cars were conceived as an assemblage of parts, a bolting together of a chassis/drive train (engineering) and a passenger coach (crafting). In those days, car designs were modeled in hammered sheet metal and wood, which reinforced the notion of an assemblage.

Harley Earl, the legendary design chief of General Motors who had developed a reputation for streamlined, fluid, custom automotive creations for Hollywood stars, introduced the industrial clay he had been using for his modeling process when he styled his first production automobile in 1927.[6] His work was characterized by a smoothing of the junctions between parts—in effect, beginning a long process of car design shifting from the "bolting together" concept to the fluid, formal concept still prevalent today. The clay modeling material provided the means: by the 1930s, full-scale modeling in clay was standard practice in major car design studios.

But did clay the *material* also provide the concept? Did Earl have in mind this smoothing quality of clay, or did the idea emerge from the material's characteristics?

To answer this question, we need to consider the means with which material is engaged. Language, drawing images, and making—three of the fundamental capabilities that distinguish our species—locate the confluence of material and mind. Drawing and making link the hand and eye to perception, just as language links the auditory organs to reasoning. This symbiosis or entanglement of making, drawing, and language comes naturally, in part because all three share common roots and traits. Whether speaking, drawing, or making something, each activity is accumulative, reflective, and reiterative. For example, making (including technical making) requires many steps, with rules and procedures formed from interactions with material. Language, too, has similar rules (syntax) and procedures that can be extended and recombined to alter and create new meaning. It is the same with drawing: each mark leads to another in a reflective way; that is, each mark influences the subsequent marks. The drawings of Harley Earl (the GM design chief) illustrate the symbiosis between ways of engaging (figure 1.1).

1.1
An early Harley Earl rendering.

Comparing Earl's drawings to his automobiles, we see an affinity between them that is deeper than the obvious resemblances. It is the drawing *technique* of the earlier motorcar that most aligns with its conception and manufacture. The drawing is carefully wrought, stippled or airbrushed to leave no gesture or hand marks. Its shapes are likely controlled with templates, compass points, and probably done over an underlay. The smoothness of the design is expressed in the attentive modeling of the curving metal, given definition by light and shade. The drawing is of a static object. In comparison, drawings from the years when clay modeling had become standard practice are highly gestural; the hand actions are looser, faster, and fluid, and lines appear as records of dynamic movements. Such presentation drawings (figure 1.2), although they weren't done as "action drawings," carefully reproduced such action gestures in form.

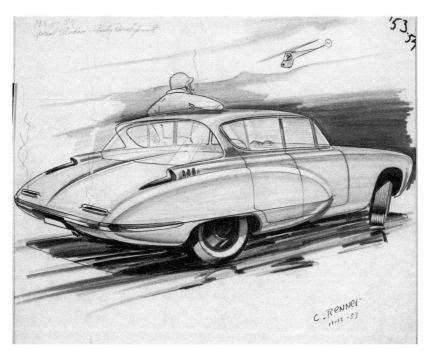

1.2
Presentation drawing, Carl Renner, 1953.

Both the clay modeling and the gestural drawing used motion to impart the sense of motion to forms. Clay modelers can be seen in historical photographs using cross-section templates to minimally control the clay, freeing them to drag swooping and curving profiles from front to rear, aligning with the drawing's strong bias of gesture from front to rear, as if to suggest that airflow was the modeling agent (figure 1.3).

Earl's 1969 obituary in the *New York Times* dramatizes how a design "decision" came from an action—not an additive one such as adding a line to a drawing, but its opposite, the kind of subtractive gesture clay modeling is all about. "In the mid-nineteen-thirties he rubbed his thumb across a sketch of a proposed new model and erased the exterior running board from the American scene."[7] How crucial this is to the story cannot be overstated. By rubbing and erasing a necessary element in car design, Earl forever changed the concept of the car.

And there it is. The symbiosis of drawing, of making, of gesturing, of thinking, conducted by and through smoothed and carved clay, smudged pencil, and arcing motions of tool and hand. If one ingredient of the

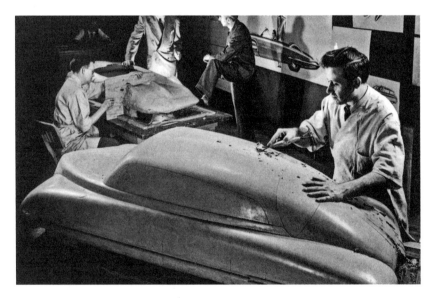

1.3
Clay modeling at General Motors, 1940s.

drawing had been different—say, indelible, permanent ink rather than soft graphite—would Earl have had the impulse to rub his thumb across the sketch? That, again, is the fundamental question this book explores: How are ideas, the imagination, and creativity influenced by and intertwined with physical material?

MIND AND BODY

When Harley Earl rubs out the exterior running board, we see his actions but cannot readily isolate his state of mind from those actions. We know, from experience, that our body, its limits, and its "material" are integrated with our minds, our thoughts. We think through the material of our body absent any self-consciousness of how our corporeal self governs us and our "free will." Only when we stop to think of the "freeness" of our will may we realize that our body's limits are implicated in our own volition; the will is bound tightly to the body. We cannot escape limits; we never trespass into the impossible ("I *will* fly!"). Any project or will to form that we project forward is simultaneously felt as free, and at the same time is manifested through and restricted by our corporeal matter. The will is necessarily felt as free to disconnect us from our limits, even if unconsciously we remain tethered.

It is this disconnection that creates the space for "maybe" and "could," aspirations, reverie, efforts, all those things that might be half-formed, half-real, that fail, that extend us. The word "will" carries an ambiguity within itself. "I will" is declarative and willful, while "I am willing" signals acceptance and compliance. Materials can be characterized by these states as well—some are willing while others seem willful, stubborn. All materials, each in its own particular fashion, meet our actions, resist them, transform them, and reflect new possibilities back at us, and in doing so bring our volition into a process of imagination and creativity.

Despite our own commonsense experience of strong interactions between body and mind, thinkers have struggled to understand the nature of those interactions. The mind/body problem has long been a subject of debate in philosophy, influencing our views and reinforcing a bias (at least in Western education and culture) toward dualistic models that see the mind as completely separate in nature from the body and physical materiality. In most dualistic models—the most famous being that of French mathematician and philosopher René Descartes (1596–1650)—the mind retains its autonomy from the body and is assumed to act on it unidirectionally.

Descartes's famous proposition *Je pense, donc je suis* ("I think, therefore I am")—which we typically see in Latin, *Cogito, ergo sum*—implies that thought precedes action and that *decisions are made through and carried forth from thought* within a hierarchical mind-to-body relationship. For Descartes, mind and matter were both substances, though of separate natures, the independent mind directing the body through the pineal gland. Although Descartes was the first to acknowledge that emotions such as passion could arise from the body and direct the mind, the popular bias associated with Cartesian duality is that the mind is distinct from and controls the body and, by extension, material (since the body is material, acting in accordance with the same laws as all other matter). This bias, so integral to our legal, institutional, and educational structures, marginalizes the physical causality behind thinking and imagining.

There are, though, propositions predicated on a *nonhierarchical, reciprocal* relationship between our senses and our thoughts that contradict the Cartesian mind/body duality and point to a far greater entanglement between physical conditions, the sensing self, and the mind. French philosopher Henri Bergson (1859–1941), in *Matter and Memory*, challenged Cartesian duality by proposing that the mind interacted with the physical world through memory.[8] In Bergson's view, the physical (which he argued is received in the mind as "image") brought forth multiple memories that would, in widening ripples of attention, meet and envelop the image (and even become indistinguishable from it), thus deepening the solidarity between the mind (spirit) and matter. For Bergson, the temporal nature of the mind is what distinguishes it from matter (which is spatial), and human action is key to mind/matter reciprocity.

Though Bergson was arguing to distinguish the mind/soul from matter, he saw them as far more intertwined than in Cartesian dualism. His familiarity with and citing of contemporaneous neurological investigations (brain pathology studies) to build his philosophical position has lent greater credence to his views, many of which have been found to fit with scientific findings.

French philosopher Maurice Merleau-Ponty (1908–1961) dispensed with the separation of mind and matter, subject and object, and viewed the body not only as central to understanding the nature of perception but also as the agent that informs and lends significance to the world while simultaneously being defined by it. He wrote in his book *Phenomenology of Perception*:

> And in so far as my hand knows hardness and softness, and my gaze knows the moon's light, it is as a certain way of linking up the phenomenon and communicating with it. Hardness and softness, roughness and smoothness, moonlight and sunlight, present themselves in our recollection, not pre-eminently as sensory contents, but as certain kinds of symbiosis, certain ways the outside has of invading us and certain ways we have of meeting this invasion.[9]

Merleau-Ponty introduced the idea of equivalence of traffic (and influence) between the self (mind) and the world (physical phenomena), an idea that entered cognitive science decades later.

"In intellectual history," wrote American philosopher and scientist Jerry Fodor, "everything happens twice, first as philosophy and then as cognitive science."[10] More than fifty years after *Phenomenology of Perception*, Andy Clark and David J. Chalmers—both, like Fodor, philosophers and cognitive scientists—demonstrated the truth of Fodor's statement when they penned their influential essay "The Extended Mind," which proposed a cognitive model along similar lines to Merleau-Ponty's.[11] The "extended mind" hypothesis defines the mind as spanning between the physical environment and the brain, arguing that there is no reason to limit the mind to the skull case.

Clark and Chalmers use the simple example of a pen to illustrate the way the brain extends itself through writing. To remember something, we write it down, and in doing so we extend our memory into the notebook we wrote in. Tools couple the mind and physical environment. If we accept this idea of the mind, then we must also accept the role the physical environment and tools play in forming thought. In other words, in extending itself into its surroundings, the mind must be directly influenced by those surroundings as if they are of the mind itself. Such a model of the mind aligns with my central thesis that working with physical material generates thought, imagination, and creative insight.

The conclusion of "The Extended Mind" suggests this model has consequences beyond cognitive science and philosophy:

> As with any reconception of ourselves, this view will have significant consequences. There are obvious consequences for philosophical views of the mind and for the methodology of research in cognitive science, but there will also be effects in the moral and social domains. It may be, for example, that in some cases interfering with someone's environment will have the same moral significance as interfering with their person. And if the view is taken seriously, certain forms of social activity might be reconceived as less akin to communication and action, and as more akin to thought.[12]

The concept of the extended mind is functionalist, involving the leveraging of cognitive abilities, such as writing notes to extend one's memory. Francisco J. Varela, Eleanor Rosch, and Evan Thompson further proposed a concept of the mind that forms and enacts itself through the active and embodied coupling with its environment.[13] Their "enactive mind" theory posits that knowledge is constructed through "enaction," that is, from the mind's ongoing process of being structurally and dynamically coupled to the environment, which includes both physical and social contexts.

In both the extended mind and the enactive mind, there is an exchange with the physical and material. The question, as I see it, is one of degree. In some instances, the extended mind is more operative; in others, it is the enactive mind. In later chapters, I argue that the nature of our surroundings and situation regulates the degree of extension and enaction of mind.

Before the advent of language and tools, the mind's immersion in the environment would have been more direct—in the way animals are completely in the moment, without the self-conscious interiority that language has given us. In my opinion, the development of language, drawing, and making (including tools) traded the immersive self for a new self-awareness that bound (extended) us to each other socially but separated us from a complete being-in-the-world. This separation is made manifest by the strong impulse we have to find significance in and give meaning to "meaningless" things, because doing so joins them to us and vice versa.

How did we get to this arrangement of self and world? The evolution of our species bears examination.

BRAIN AND HANDS

What distinguish our species zoologically from our cousins are seemingly minor traits, given the diversity of the animal kingdom. Our entire enterprise of "humanness" has managed to build itself on a thimbleful of the byproducts of bipedalism, the most critical of which is the differentiation of "intelligent" hands; freed from locomotion, the hands developed new capabilities firmly connected to and preceding—and thus implicated in—a subsequent increase in brain capacity and changes in cranial morphology. "Lucy," our famous upright-walking *Australopithecus* ancestor from 3 million years ago, had a very small brain, but she was already irreversibly committed to the path of toolmaking as an extension of her biological limits.

All evolutionary changes involve tradeoffs. While the possibilities seem infinite, they are bounded by limits of strength, energy use, physics, and mechanical forces, among other things. The tradeoffs of our ancestral line are telling. Walking upright transformed our skull; as the head became more supported from below, the jaw and teeth shortened to make room—a limit anyone who has had wisdom teeth removed can appreciate. With a smaller jaw and smaller teeth, the face no longer retained its primary role of grasping and acquiring nourishment for survival. First this diminishing capacity was subsidized by the hands, and eventually the stone biface chopper tool completely replaced quickly disappearing incisors. With our face having become ineffectual for grasping and cutting, our species became completely reliant on tools for survival.

André Leroi-Gourhan (1911–1986), a French anthropologist and paleontologist, traced an arc of human evolution in his classic *Gesture and*

Speech that links the anatomical to our technical, social, and symbolic impulses.[14] Our species is the only one that made the leap to externalizing and extending its genetically determined limits. The deep impetus of technology relentlessly externalizes our physiology: muscles are extended by levers, wind power, motors; eyes by lenses, telescopes, cameras; so, too, the brain, first with valves, then electronics, and now computers. Our social organism, caught up in and dependent on technology, exults in its liberation from biological restraints, while our zoological organism remains stuck in the slow lane of evolution. As our technosocial organism evolves at an ever-increasing pace, the gap between physiological limits and technical capabilities has become immense.

Before technology began to accelerate the human project, there was a period of several million years during which technology barely advanced. Stone tools were unchanging even as new species evolved, first *Homo habilis* followed by *Homo erectus*. Similarly the basic anatomical "design" was already in place with Lucy; subsequent major changes happened in the head. Erect posture and smaller jaws created extra room in the cranium into which the brain began to grow, from Lucy's 400 cubic centimeters to the 650 of *Homo habilis* and the 900 of *Homo erectus*. The arrangement of the evolving brain was such that new areas controlling speech and touch developed proximal to growing hand motor-control areas—concrete evidence of the interdependence of our face and hands. Given that the hands are so deeply integrated into the centers of our intelligence, their use constitutes a large portion of our mind.

Animals' motor reactions to sensory input are instinctive, direct, and instantaneous, with fewer resources devoted to reflection and the consideration of multiple courses of action. The areas controlling such "executive functions" grew significantly in hominids, and the prefrontal cortex—as it is called—began to impose itself strongly between motor and emotional functions, suppressing, stimulating, and integrating them. This arrangement defines human intelligence, the tradeoff being the radical alteration of our being-in-the-world. Our growing self-awareness has put the world at a distance. We can contemplate it without participating in it, but in doing so we have lost the seamlessness of being—what athletes and performers strive to recreate when they speak of "being in the moment."

Homo sapiens emerged from millions of years of infinitesimally slow progress with a new brain, one that inhabited its body, conscious of the gap separating the mind and material world. It was so then and remains so today: our detached frame of mind tends to think beyond the limits of the body, making imagination and ideas possible; at the same time, the limits of our body and the material that surrounds us are the activating agents of and subconscious contributors to the creative mind, a mind that continuously convinces itself of its self-determining sovereignty.

The last ten thousand years are a triumphant demonstration of our ability to build an artificial reality out of the natural world, parallel to the project of externalizing the physiological body into the technosocial organism. Naturally, it is the aloof mind that, despite its evolutionary intertwined codevelopment with physiology, has continued to subordinate the hand throughout history. Even the earliest of civilizations discriminated against those who created with their hands: whereas priests were holy, rulers were brilliant, and judges were wise, artisans were merely skillful. Despite having made and in large part invented the new world, artisans were habitually derogated as an underclass somewhere above slaves and below merchants.

As the artificial has continued to dominate, the natural and the hand have remained subordinate to the head, and we continue to exert command and control over the limits and interdependences we confront. Human advancement, especially technological advancement, never escapes its physical contingencies, but with each passing decade our species persuades itself it has done that very thing.

ENGAGING MATERIAL

"Meaning," archaeologist Lambros Malafouris has written, "is the temporally emergent property of material engagement, the ongoing blending between the mental and the physical. In the case of material signs, we do not read meaningful symbols; we meaningfully engage meaningless symbols."[15]

I wrote earlier that our engagement with materials goes beyond physical manipulation; images and language are also forms of engagement—what we say or draw about a material affects our relationship with it. These forms of engagement aren't separate but symbiotic—words guide making as do images, and in turn images are informed by words and vice versa.

This commonality of structure and process is consistent with the brain's architecture. Neurologists used to think the particular region of the human brain that processes language syntax—Broca's area—evolved exclusively for that purpose. Recent research demonstrates, however, that making activity is also controlled by Broca's area, which raises the probability that Broca's area evolved first from toolmaking and only later was coopted for language.[16]

Making, images, and language define what we *do* to material (figure 1.4). To see what material *does* to the mind requires shifting attention away from material to where the mind intersects with material. Let's say we squeeze a lump of clay. It squishes somewhat unpredictably about in our grip, but doesn't affect our thought. But if we decide to make the clay into something such as a brick, a cup, or a tile, we are inspired—and this comes directly from the clay's material behavior—to act, not with more squeezing but with *tools*.

Tools, in this sense, attempt to join intention and material behavior, and in doing so define a meeting of the mind with material. Tools can

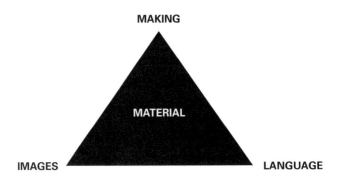

1.4
Material's relationship with making, images, and language.

best be understood as operating between the categories of action (making/images/words). Between words and images one finds symbols and signs (figure 1.5), which are tools for extending the mind and making material meaningful, the notches in the Ishango bone being one such example. Between language and making one finds all the tools that combine the procedures, assemblies, and rules of language with the demands and limits of physical material. The rock-paper-scissors game distills such a condition: logic and material properties are exchanged, one leading to the other, as paper lends itself to being cut by scissors or, conversely, to covering a rock. Located between making and images is the newest arena for tools, which encompasses all tools used to plan out the making of something by first developing it in two dimensions with the use of technical images and digital means. Such disembodied translation between two and three dimensions has been with us for several hundred years, but is often overlooked as simply a technical means to an end. These tools, such as orthographic and perspective projection and digital scripting and rendering, have had a

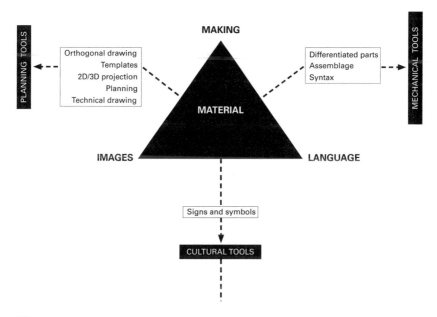

1.5
The tools of material's relationship with making, images, and language.

much greater impact on our relationship with materials than is generally realized, and discussing them is key to understanding how we have convinced ourselves of our absolute mastery over our material world.

These fields of operation—planning, meaning, and procedures—come into play through actions and are of special significance, situated in the territory of overlap between mental operations and physical/material actions. This "common ground" between material and the mind is *the* fertile ground on which to form concepts from material and on which material gains significance. I would argue that it is on this ground that some of our most original and creative thoughts and ideas emerge, and it is here that we can find insights into how and why we do and make.

Despite a general indifference to material's role in thought processes, this doesn't mean there isn't a consciousness of it. Material is "mapped" in our mind such that we can discern all sorts of properties from the "model" of wood we have in mind—a combination of experiences and assumptions. In addition, material is profusely appropriated by thought—as metaphor— and we rely on our experience of material to help express immaterial concepts: heart of stone, sands of time, earthy honesty. But words alone aren't actions, and it is actions that disrupt thought's sense of detachment. When we manipulate actual material, wood for instance, it continually surprises and overturns the mental "model" of wood—which is why the wisest of craftsmen are so respectful of the material they have mastered.

Perhaps, though, to understand the answer to the fundamental question of how ideas, the imagination, and creativity are influenced by and intertwined with physical material requires posing a precursory question: From whence—regardless of the role of materials in it—does the thought process spring?

DOUBT

"What do you mean you're not sure?!" the young student exclaimed. "You're supposed to be the teacher!"

She was clearly upset. In her mind, the elaborate structures of life, the scaffolding that undergirds our everyday actions and assumptions (including those of the classroom), had to remain stable, humming in the background, for her to do "the assignment." I was teaching first-year college students in an art and design foundation class and nudging them to consider their work less as "fulfilling a requirement" and more as an iterative process that would construct itself into a proposition.

"I'm not sure," was my response to her question. "How *do* we know what to do next?"

I was struck by her insistence on having a plan of action before proceeding, a willingness to forsake imaginative and unexpected potentials to maintain predictability. In such a class, full of young artists and designers, one might expect a natural affinity for action, for experimentation to see or discover (imagine trying to make a painting simply by planning it all out in your mind). But that wasn't (and still isn't) the case. Education has long sidelined playing as a means of learning, and my students—coming to my class after a dozen years of being taught to the test and pressured to achieve grades—had naturally developed strong tactical planning habits aimed at "completing the work." They had become academic "achievers," products of a system that rewards knowledge more than understanding or questioning. To them, doubt—in the form of unsureness—was anathema.

What about *Cogito, ergo sum*, then? And what about that reciprocity between thoughts and actions?

Research on neurons firing in our brains has demonstrated that "thought" may arise long after action is already under way.[17] For instance, when facing danger, we *act* long before we're even aware enough to *think*. This may be difficult to accept: it's like finding out that our inner, essential, timeless "I" is more puppet than puppet master. Recent discoveries in neuroscience are casting doubt on the notion of "free will," the immaterial soul, or a self that exists independently of the physical body.[18] Despite this new evidence, we continue to live in an ethically and morally constructed universe based on personal responsibility—one in which our actions are fully our own, stemming from an autonomous self. That is what our self tells us, after all. And that self is persistent: notwithstanding the Copernican revolution, we still see the sun move from east to west, and it's rather difficult to "feel" the earth rotating from west to east against an unmoving sun.

As the American philosopher, psychologist, and educational reformer John Dewey (1859–1952) wrote, "To be genuinely thoughtful, we must be willing to sustain and protract that state of doubt which is the stimulus to thorough enquiry, so as not to accept an idea or make a positive assertion of a belief, until justifying reasons have been found."[19]

Cogito, ergo sum is actually preceded in Descartes's work by some words of explanation for why we think. In an essay praising Descartes,[20] French literary critic Antoine Léonard Thomas (1732–1785) offered a fuller version of Descartes's proposition, based on this explanation: *Dubito, ergo cogito, ergo sum*, or as Thomas originally wrote, *Puisque je doute, je pense; puisque je pense, j'existe.* It means "I doubt, therefore I think, therefore I am."

Doubt is as much sensation as a part of reasoning, and in both cases an unstable state. We all know the bodily sensation of doubt when queasiness, a tensing or weakening of the musculature, courses through us. With its origin in "fear"—the Latin word *dubito*, meaning "I waver" or "I hesitate," made its way to us through Old French *douter*, or "be afraid"—*doubt* (our English word) precedes thought and even overrides it, leading us to action because doubt is such an uncomfortable sensation: *Dubito, ergo faciam, ergo sum*; "I doubt, therefore I act (or make), therefore I am."

MEANING AND EMOTION

Our common sense would seem to tell us that the kind of "doubt" we have about ideas, or opinions, or people—cerebral doubt—is uncoupled from physical sensation, a "state of mind" above all this bodily nonsense. But emotion plays a part here as well: the joy of solving a puzzle is no less a physical sensation than that of an embrace. The state of doubt John Dewey wished us to sustain is challenging because doubt is negatively felt even when we are "only" thinking. There's a powerful human tendency to avoid doubt. It is a tendency so strong, as political science researchers have discovered, that belief in wrong opinions is unaffected by factual corrections and even strengthened when confronted with such information.[21]

"Belief," too, is rooted in emotion; the word comes to us from an Old English word, *geleafa*, that itself meant "care" and even "love." Just as with the *Omnia vincit amor* ("Love conquers all") of Roman poet Virgil (70 BCE–19 BCE),[22] deeply held beliefs provide an unshakable foundation for an outlook free of doubt—one that is also unthinking and unimaginative. To have doubt is to be in a vulnerable position, uncomfortable, unsafe, and potentially threatened. But for anyone trying to discover, create, imagine, or feel, doubt marks the territory of operations. It gives us access to perceptions, interactions of the senses and the mind, or as German philosopher Immanuel Kant (1724–1804) might have defined it, aesthetics.[23]

We tend to impose concepts of uniformity and stasis on materials in order to stabilize our physical surroundings. When a material doesn't fit our mind's assumptions—and it never does when we engage it—feelings such as doubt and belief emerge to be important ways to navigate the gulf between the mind's stable concepts of material and material's mutability.

Today, the question of how our physical, material environment informs, forms, and guides our thoughts and insights has greater urgency as we approach the limits of an acquisitive and detached intellectual framework. As long as we regard the means with which we act on the world as simply propagative, we will remain blind to the intelligence coming back through our senses and through the material world surrounding us. For the student who so badly wanted a plan of action, it took a long time, months and years of working with materials, to realize that receptivity and attention are both key to any creative process.

If we keep our mental receptivities and sense receptivities open, doubt can unify them through the language of emotion. Doubt, then, deeply entangles thought with feeling, so deeply as to suggest thinking arises from physical/material conditions received by the senses, as much as our apperceptions are shaped by thought.

This is perhaps nowhere more apparent than in how we imagine.

2 IN THE BEGINNING

We place a high value on the imagination. It defines our species and advances human endeavors, from culture to science. How imagination works, where it comes from, and what activates our imagination should be of interest to everyone.

In subsequent chapters, I discuss in greater detail the imagination and the questions that surround it. In this one I want to explore how the imagination may have originated and whether it exists only in the mind or is part of the enactive process or extended mind—that is, whether it is not only linked to thinking but also originates in *doing*.

The dictionary defines imagination as "the act or power of forming a mental image of something not present to the senses or never before wholly perceived in reality."[1] More exactly, imagination can be sensory, giving us the ability to make an image in the way our senses (mainly visual) perceive, or it can be cognitive, giving us the ability to form ideas or possibilities. For Aristotle, the imagination spanned between the two, between mental images and ideas, combining them in the act of reasoning—all within the space of the mind.[2]

The idea of a private and internalized imagination is persistent in Western thought, seen always as operating in the space between perception and dreams or the fictive, but never outside the mind. That is, in a sense, the *raison d'être* for the imagination—it is the mental faculty that mediates the disjunctions between externalized, material reality and the internalized mind. In other words, the imagination exists *because* our mental conceptions do not correspond with the reality of things. This has led to an indictment of the imagination, from the Enlightenment on, as a symptom of inadequate powers of reasoning or, at best, a kind of fanciful "artistic" activity. I wish to reestablish the imagination's role in bringing images and ideas together, and to propose that its activation and power depend far more on the body and, by extension, on physical material than is usually assumed.

Henri Bergson, in *Matter and Memory*, saw time as key to understanding the relationship between matter and the mind. In his view, the body and its perceptions and sensations are more a part of material things than of the mind. The mind links to the body's perceptions (and thus to matter) through memory (and imagination). Bergson saw memory as triggered by and guided to perceptions, overlaying them so precisely that "we cannot say where perception ends or where memory begins."[3] This is a way of saying that memory and imagination are drawn to the present by the present—by the body's active perceptions of matter. Memory and imagination are thus embodied in and affected by matter.

Bergson further differentiated the degree of autonomy of what he called "pure" memory from that of pure perception along a temporal scale: the plane of impulsive action that is closest temporally to perception is more about conscious, even automatic association, while the plane of dreaming that lies furthest temporally from the present (and perception) gives rise to autonomous representations, mental images that are most disengaged from immediate perceptions. Bergson concluded that the ideas and images of memory and the imagination are formed by a roving mental activity of the mind oscillating between action and representation, along a temporal scale. Such a model of memory and the imagination bridging between the inner self and physical reality, between reflection and action, comports with the idea of the extended and enacted mind—in this view a continual unbroken circuit from material to perception to reflection/imagination/action and back to material.

In this chapter, I show how these ideas of the imagination, as enactive and arising from the physical, are part of everyday commonsense experience, despite the persistent notion of an autonomous imagination sequestered in the mind. Let's begin by examining the differences and similarities between *imagining* an image and *making* an image and how physical reality is always part of either picture.

Our belief in our ability to conjure or precisely recall a mental image is so ingrained that most of us give little thought to what is really going on. But if we test ourselves—something I've done with my students over many years—we find that our ability to examine the contents of a mental image is far more limited than we may have thought.

You can test this yourself. Try picturing a building you've walked by repeatedly over the past few years. Can you recall its details? How many windows does the building have? Are there lights by the doorway? Where? What are the materials from which the building is made? Can you mentally assemble its parts? All such questions can be answered quickly by looking at a photograph, but your mental image will surely lack such specificity. This is one reason eyewitness accounts are notoriously prone to distortion; even witnesses' recollections of criminals aided by artist's sketches are mostly too inaccurate to identify actual individuals.[4] Words, too, can alter a mental image: consider how many times a courtroom lawyer has demonstrated this fact by using suggestive language.

We cannot know whether mental images came to be because of drawing's influence on perception or vice versa, but we can see that mental images share properties with drawing, namely, their high degree of condensation of visual information. Even the most elaborate drawing is a drastic distillation or reduction of the information that you might get, say, from looking out a window. Such a reduction of information allows mental images to be stored and retrieved easily in the mind. And drawing has a further connection to mental images in that the act of drawing can somewhat enhance or help reconstitute a mental image—although this is hardly a linear or precise process.

As limited as a mental image is, it still has great utility in the drawing process. It is the quintessential visual aid. It allows us to "hold on" to what we last saw as we shift our attention to the surface on which we are drawing. More significantly, the constrained bandwidth of a mental image shares the reductive and abstracted conditions of, say, a line drawing—so much so that it hints at a mutual origin: the act of making marks having led to the mind's capacity for the visual memory of those marks, which in turn led to a greater capacity to build marks into an image.

The intimation is that of a mind learning how to perceive and remember from the act of drawing.

AWARENESS

American psychologist Jerome Bruner (1915–2016) defined "meaningful" as having a sense of connection and relevance that is greater than self.[5] The impulse to mark or remember with a mark is one of many ways we build a deeper connection to place and, in doing so, make the world around us "meaningful." Footprints in the snow or mud are inadvertent records of passing by; it's a small mental step to make such imprints purposefully, to mark one's presence by declaring, with an imprint, "I am here" or "I was here." A handprint on a wall, commonly found in early cave art, fulfills such a role, with an added level of identity because our hands are more bound to our selfhood and self-expression than our feet. But when early humans placed their hands on cave walls and blew charcoal dust or pigment over them to create outlines, it was a remarkable leap from purposeful imprints. It opened new conceptual territory. Like the gestures and grunts that evolved into language, the dust-blown reverse image was no longer mimetic in the way a footprint is to a foot or even a handprint is to a footprint (mimetic technique). It stood alone as *something else*.

The *techne*—the *art*—of blowing pigment has had unanticipated outcomes, most of which are far more potent than the original goal of making a handprint. For instance, this technique, akin to stenciling, gives the maker the power to reverse and control solid-void relations (or so-called "positive" and "negative" space), conferring authorship to the process. By spraying both light and dark pigments, Paleolithic artists created the impression of depth—the first abstract representation of the third dimension—35,000 years before projection geometry was formalized. While handprints and stencils are parallel developments—one is not more "advanced" than the other—the stencil technique brought traits to light that may have deepened the sense of relevance and connection for its authors.

In one of the most touching examples of Paleolithic stenciling, the Cueva de las Manos in Argentina, the proto-space of communal coherence and order appears in the overlaying of stencils and merging of images. It is a layered embrace of commingled hands in space and time. The overlapping of stencils in principle would make it easy for one to obliterate or ignore an underlying image, but in the Cueva de las Manos the exact opposite has taken place: each stencil connects to another, and some images intersect others, but the net effect is of a school or flock. A deeply connective sense forms the work into a whole—not a composition but more pattern- or fabric-like, an unselfconscious expression of the ties that bind each "I" to others—and, by virtue of the pigment and technique, melds the hands with the rock surface and the world. Such a story in place and time, a union of individuals, family, friends, traveling comrades, and tribe, is the story of life. It is how we bring measure and render visible the immeasurable subjective matter of our lives, our feelings, dreams, and laments in a medium of our shared actions. In doing so, life is made meaningful—deepening the sense of connection and relevance that is greater than ourselves.

2.1
Cueva de las Manos.

To conclude that there is a causal relation between stenciling and its purpose, or even to posit that one begat the other, erroneously assumes a passive propagative role for technique—that intent is "carried out" by technical means. In truth, the relationship between technique and intent, like the play between thought and action, is an entangled one, a hopeless knotting of fishing lines. Insisting on the primacy of one over the other achieves as much as pulling brutally on the ends of the lines to undo the knot.

Might an examination of what available technique *is not* being used yield some clue to the relationship between technique and intent?

The Paleolithic technique of stenciling had unintended possibilities, the most obvious to modern eyes being the anticipation of projection, or systematic transmission of visual information. But for whatever reasons, our early ancestors never used light to project shadows, a technique conceptually proximate to the projected pigment creating hand "shadows"—or if they did, no evidence has been found. Those early artists built fires in the deep caves they painted and saw their shadows flicker in sharp relief as they made images. Why not make a tracing of the shadow outline—just as the stencils did? Could the physicality of the blown pigment have been more rite than *techne*, more process and action than means to an end? If so, it is conceivable that the tracing of shadows would be a foreign act; if it remained culturally unendorsed, tracing's utility would be unknown. But perhaps, more simply, the explanation lies in the fact that a handprint is an enactive materializing act, a binding of the self and the world, while a traced shadow is a distancing and dematerializing act, a *displacement* and abstraction of the self. After all, images introduce distance.

The first stirrings of internalized self-awareness (in which making images would have played a part) must have been accompanied by an imagined, stress-filled future, in turn evoking a deep need for reassurances and predictability—precursors to the elaborate structures we have developed to insulate ourselves from the angst of life: narratives, rites and rituals, rules and constants, physical and spatial order. "Religion," from the Latin *religare* ("to bind" or "to tie"; by this derivation, our English word "religion" is also linked to the Latin *ligamentum*, meaning "ligament"—the fibrous connective tissue that connects bone to bone), may have originated as a way back to the felt totality, the unmediated "in the moment" inhabited by the preconscious self. The paradox of religion lies in its focus on acts, devices, and representations to narrate an alternative future in life and after, as part of the quest "back," not to the here and now but to an idealized deity—whether a god or gods or animal spirit.

Hunting in those times also speaks to awareness. It has always figured into the discussion of cave drawings, and it has to be understood in context. Hunting was not only essential to life, but success in hunting required that hunters get very, very close to their quarry. This depended on an exquisitely intimate, empathetic, even sacred understanding of the hunted by the hunter. Cave drawings are thought to be not so much visualizations of hunts as an ecstatic drawing out and touching of animal spirits dwelling within the rock. The oldest such drawings, found in Sulawesi, Indonesia, date to about 40,000 years ago.[6] There, hand stencils coexist with drawings of animals.

Jean Clottes and David Lewis-Williams, experts in cave paintings, have speculated that the cave wall was conceptually a membrane between the human and spirit worlds.[7] Even though today we may look at cave paintings as abstracted, "projected" (from memory and mind) representations of animals, the cave wall as membrane would suggest an inversion of this: that the artist, in a hallucinatory state, has brought the animal spirit from within the rock into contact with the membrane, and just as with the hand stencil, has followed a process (a rite?) to "record" the touching of spirit and human on the rock wall membrane as image. The marked membrane fuses subject with object, collapses the dangerous distance between animal spirit and human—a dangerous distance since, once the animal is "at a distance," it is no longer available to provide for the tribe's survival. In this reality, touching the membrane with hands and casting hand imprints with blown pigment is a brilliant, wholly effective technique that fuses the self with the other at the membrane. In such a magic enactive (in the cognitive sense) moment, the image may appear to be simultaneously in the mind and on the membrane, a closing of distance.

Parenthetically, no Paleolithic drawing or painting employed foreshortening (portraying an object as closer than it is or having less depth or distance), size diminishment, or perspectival convergence; such suggestions of spatial depth would be completely at odds with the notion of animal spirit and human as "collapsed against the membrane."[8]

As the image in mind fades, the membrane surface of the cave holds a memory. This differs from oral memory, which must be passed on and which changes with each iteration. A fixed memory image, which can be understood by others, has the potential to revolutionize social organization, allowing much larger, more disparate groups to cohere as a culture. In that sense, these images from Paleolithic times function like text—long before the advent of written language.

To our modern eyes, the stencils can be understood as a kind of graffiti—an enacting mark, a process—but the animal image is harder to see in the same way. The drawing tempts one to consider it a result of an intention, that is, as an image begun as "thought" that was then executed. Those who draw know that the meaning and significance of drawing is formed in its process of enactment, just as rite and/or ritual made stenciling a significant act. Drawing (especially drawing from memory) is an action within which realizations are made: making a mark prompts a reactive mark and, as marks accumulate, an image develops. It should be noted that the image cannot be fully anticipated, because the final result is contingent on the interaction of the hand and the medium and material it is made with and on. As an image emerges, the process of its making inevitably falls away. That transition from action to representation reinforces a misconception of images as intentional representations of a mental "picture."

It is noteworthy that many Paleolithic drawings incorporated underlying surface features of the unprepared rock into the drawing.[9] Could this point to the origin of drawing as little more than the graphic enhancement of an existing feature on the surface? Such an act—say, turning a bump in the rock into an eye—lies at the heart of the imagination and representation: "seeing" or imagining a different image when looking at a fixed physical condition. Here the triggers of the imagination include the bump, the artist's actions, and the drawing's image.

To make an image, our Paleolithic artist's perception had to navigate between the flatness of the surface drawn on and the implied depth and transparency an image produces (we look past and through a surface when looking at an image, making that surface transparent to our mind). There were also the contradictions of seeing and representing; for example, a drawn line exists both on the drawing surface and "on" the subject being depicted. This has to be understood while learning to distill and abstract visual perception enough to "see" the animal as "it" emerged from semi-controlled marks made on the wall. Surveying thousands of Paleolithic drawings, the scholar John Halverson noted that they share a number of attributes such as consisting of profile-view outlines that economically depict salient features of animal types, and a tendency to leave figures open, as incomplete works in progress—for instance, by missing or only minimally suggesting legs or feet.[10] Such incompleteness merges the conceptual and recognitive with the perceptual. To me, this defines the imagination not as a capacity to produce mental images, but rather as an ability to "see" more than should be possible—which is an ability contingent on a specific context.

Every drawing has a triggering moment when initial marks coagulate into a deep sense of connection between subject matter and image. In drawing from observation, for instance, that moment could simply be the next instant after an abstract curve of a line is laid down and it is recognized as the perfect repose of a viewed elbow and arm. When that happens to me during the process of drawing, the sensation is one of calmness and an uncanny synthesis of perception, subject, and mind. This reveals the imagination as a form of bridging, connecting between perceived phenomena and representational intent. When the imagination connects the two, it is as if the mind reinhabits the world, now self-consciously.

I believe this is key to a deeper understanding of the imagination as part of a process that brings two unrelated things close enough together for a spark (of imagination) to join them, which is as true today as it was for Paleolithic artists. The critical point is that the imagination depends on the near alignment of things and is not capable of boundless leaps to conjure mental images freely in the mind. Imagination is *contingent*.

Our Paleolithic artist may not have thought of drawing as described above, but the locus of drawing in either case is somewhere between mental conception and reality, simultaneously constituted of both and neither. But what do we know of drawing's specific interactions with the mind? Those who have traveled and spent time drawing new scenes *in situ* are familiar with the process: drawing forces one to slow down or stop, to focus and pay attention over a considerable length of time—much longer than when taking a casual photograph. Unlike photographs, which are difficult to remember until they are looked at, drawings persist in the mind as some kind of mental image. The act of drawing creates these images and infuses our memory with them. Without looking at my sketchbook, I can still trace the outline of an Italian landscape I sketched a decade ago. The memory I hold is not only visual but of the heat of the day, the light, silence barely broken by occasional solitary birdsong. In the act of forming a "mental image," the definition of imagination offered earlier also defines this sort of memory of a drawing, a memory that, like imagination, proliferates associations. Perhaps the only difference is the direction in which they point: memory to the past and imagination to the future.

Those who draw also know that the marks of a drawing do not duplicate in any way the phenomena of perception. The French philosopher Maurice Merleau-Ponty (1908–1961) explained this in his essay "Eye and Mind," using the example of an apple outlined by a continuous line.[11] The outline seems simple enough—it follows the edge of the apple. But at what depth is that line in the space of that image? The line belongs neither to the near edge of the apple nor to the background behind; rather, it belongs to both simultaneously. Artists know this: the line is impossible; it isn't in reality; it can't be here and back there at once; and thus the line exists only in drawing. Lines are marks that do not duplicate reality but nevertheless can evince reality. Purposeful or not, acts of marking, actions on materials, would have preceded drawing (like a child's scribbling having no intent). That a line became associated with image—that is, that we relate a line to a reality—demonstrates how much the mind hungers to be immersed in the world and how perception is as much an enacting sense as it is a receiving sense.

An image not only has the capacity to serve as an external memory, but also brings self-consciousness to the act of seeing. Archaeologist Lambros

Malafouris speculates that the origin of image making provided "a scaffolding device that enables human perception to become aware of itself."[12] He uses "scaffolding" to suggest perception learned about itself only when "caught" acting outside its interior cognitive state, caught as it was constructing an image. To the scaffolding metaphor I might add "net," which shares the deformability of a membrane and the geometric framework of scaffolding. All three metaphors—membrane, scaffolding, and net—are exterior to the mind, but are structures the mind can perform with and on. An image catches the mind's entanglements with the world through bodily enactment, and unlike any other human activity the image reveals the structure of perception.

Malafouris, as well as Clottes and Lewis-Williams, sense what is true of all images: like the artifice of a drawn line that exists here and there simultaneously, an image is different from visual experience in space. It is a surface that can be looked *at* and looked *through* simultaneously. When you are "seeing" the wild boar on the cave wall, you are no longer looking *at* the wall's surface but *through* it to the boar.

This trick of perception is bound to action and movement. If you walk about, you perceive the world as almost opaque, and even images you may pass by lose their "transparency." Sitting and facing a surface, however, changes perception. Images become more readable, that is, transparent and representational. The physiology of vision supports this: most visual processes are body-centered, making our actions and perception highly interdependent.[13] Perception, action, and conception work together similarly in image making: perception's fluidity and changeability, coupled with the stability of intent and preconception, are alternately motivation and restraint for each other, governing actions. Making an image is a symbiosis of perception, action, and intent, with no single one of those taking precedence. It is a virtuous circle that leads to a result not imagined in the mind but created over time, through action.

For the early artists, the making of an image broke the unselfconscious continuity between self and all. Like trying to see into a mental image, drawing disrupts the illusion of seamlessness between what our eyes receive and what our brains visually think. To draw is to reveal how the mind forms visual information into an internalized schema of perception that does not align with ocular information. For example, our brains tend to perceive similar objects receding into the distance as being the same size even though our eyes do not receive the information that way, which helps explain why it is fundamentally unnatural and difficult to draw in perspective. To draw poorly isn't a problem of eye-hand coordination, or of mastery of a medium, but rather an inability to mediate between what the eye sees and what the brain tells us we see.[14] The act of drawing becomes the enactment of perceiving the world as it is, even as the marks of a drawing and its two-dimensional nature create an artifact that can never be confused with reality.

BEYOND SEEING

Animals are said to stop the hunt when their prey disappears. It is thought they lack a persistent visual memory—the ability to "see" something after it is no longer there or, for that matter, to see something that doesn't exist—which is the foundation of the imagination.[15] If early hominid hunters had also lacked a long-term visual memory, they, too, would have been more intensely "in the moment" without the distraction of memories or daydreams. The implications for survival strategies that accompanied the development of bipeds are worth considering, since the slower speed of bipedalism was a disadvantage for capturing prey. This relative slowness was offset in part by the hunter's ability to run over very long distances, and more importantly to persist intensely in the hunt. Without a strong visual memory, given changing topography, land features, and vegetation, prey would certainly "disappear" long before the hunter's endurance ran out. In this scenario, the development of a deep visual memory could have been an advantageous response to survival pressures, allowing a hunter to exploit fully his endurance by being able to continue seeking a prey periodically lost to sight.

Image making requires a strong visual memory—the ability to see what is not present or is only partially suggested. Image making is a cultural activity, but this doesn't mean it emerged as such. As paleontologist and evolutionary biologist Stephen Jay Gould (1941–2002) explained, "*current utility* bears no necessary relationship with *historical* origin."[16] Evolutionary changes can occur only if there are selection pressures and genetically carried choices from which to select. We cannot say that visual memory preceded drawing or toolmaking or even that visual memory was selected as a response to survival pressures. We can only sketch out such a possibility.

If a greater visual memory provided an advantage for long-distance hunters to leverage their powerful persistence and track quarry as the hunt went on, it may have come at the cost of losing that singular and hyperfocused "in the moment" modality. A strong visual memory, after all, allows us to reflect on the past, to contemplate the nonexistent, to separate our "I" from an unselfconscious and totalizing present. The great emotional consequences of visually remembering and imagining—who hasn't felt the intense joy of anticipation, even more intensely than the joy of the moment?—further distracts from the here and now.

If Paleolithic drawing was a "drawing close" motivated by the urge to collapse distance, the action of drawing and painting also awoke projection (think of rays depicted as lines). When a pigment-covered hand was *lifted* off its stenciled surface, it offered a first glimpse into a new possibility. The space between hand and its stenciled charcoal image was the space of projection (such as the gap between an object and its shadow), an immense, untapped arena ready to activate image, memory, and emotion. Would this proto-concept of projection be useful to the rock artist seeking to fuse with the all? Projection "throws" at a distance, separates subject from object, and constructs space between a *contemplating* audience and subject/object. Projection is inseparable from representation: what is projected is never the whole, the totality, but is always some aspect, some re-presentation of a part.

When light was first used to project a shadow on a wall, the touch of the hand was no longer intimately tied to its image. The projected shadow carried with it the capacity for disembodied action (such as turning a form to change the shadow/image) and an important new uncertainty: shadows get blurry, move about, and are more difficult to draw than blowing dust on a hand. This approximate condition of drawing must have been fertile ground for all sorts of "mistakes" of judgment, motor control, and instrument design. We do not know enough to discern the play between intent and enactment, but surely the approximate nature of drawing images— even if not consciously projected—opened the space of projection to the realm of the imagination and the interior world of the self. Early drawings, like our contemporary loose sketches, must have conjured associations, the way inkblots or clouds do, as they edged from instrumental trace toward autonomous representation.

And from there it was only a small step to drawing the imagined, the unseen.

MAKING AND SPEAKING

How does *making* figure into the interstices of language and visual thinking?

Paleolithic artists, whether hallucinating or not (a possibility alluded to by Clottes and Lewis-Williams), had to have developed some capacity for visualization (visual memory) to draw what they drew, since they were not working directly from observation. One group of anthropology and neuroscience researchers concluded that human brains expanded rapidly, tripling in size, between the advent of simple (Oldowan) tools 2.5 million years ago and more refined (Acheulean) ones of 250,000 years ago.[17] Using PET scans of toolmaking volunteers, they detected that toolmaking activates both visuomotor and language circuits of the brain, suggesting a connection between visual art, toolmaking, and language. Given the assumption that survival strategies drive evolution (such as a longer neck meeting a need for increased access to forage) and that elective activities (neck rubbing) are simply consequential, toolmaking (technology) is typically regarded as the indexical activity to mark the evolution of our species.

However, if visual art, toolmaking, and language coevolved, as research suggests, it implies that each influenced the other's evolution. In other words, the development of visual art influenced the development of toolmaking and language as much as toolmaking led to the development of art.

Visuomotor areas are intimately associated with language-processing ones; as we wave our hands while speaking, we are demonstrating the shared neurological legacy of manual action and speech. Psychologist Michael Corballis has argued that language evolved from increasingly elaborate gesturing and, in fact, that language's very structure was found within the efforts to communicate through gesturing. Another evolutionary speculation is that the technical requirements of stone toolmaking—specifically, the planning and execution of hierarchical, sequential operations—were those of language as well. The overlap and entanglement of phonological and manual control circuits raises the possibility that articulation control, a key part of language evolution, may also have figured in the development of human toolmaking.[18] It is possible that stone tools were "spoken" into being and language was "made" by manual action.

The repeated operations of stone toolmaking and bodily absorption of technique through practice are not heritable traits but were nevertheless passed on; the archaeological record provides ample evidence of this in the remarkable consistency of stone tools over long periods. In other words, toolmaking was not some kind of one-off experiment by individuals, but rather an important cultural activity that must have been encoded and conveyed in part by proto-language and mimetic representation.

Making and visual memory/imagination have perhaps the most direct relation. It seems obvious that making something requires both visual judgment and the ability to "see" an outcome. Studies indicate that the evolution of stone tools corresponds to increased visual capabilities; later, toolmaking is marked by symmetry, which meant that toolmakers needed to remember visually the shape and flake patterns of the nonvisible face as they knapped (chipped) flakes from the visible side. Anyone trying to make two sides of something symmetrical—sides that cannot be seen simultaneously—can understand how visually challenging that is and how such an achievement might be a milestone in evolving visualization.[19] Stone toolmaking was highly procedural and, as it evolved, required more planning and stages of operations and dexterity—the same sorts of mental operations as in image making. And in both cases, mastery would include the ability to envision the outcome ahead of the action—to see the flint chipping *before* it is hit, to see the form *before* it is drawn. Such envisioning is not something separate, preceding and imposing itself on form and material; it occurs in the moments of the action. It is experienced as an expression of projected confidence, less as an image and more as a knowing sense of the cognitive map unfolding in the moment.

We reflexively accept the premise that we can "see" before our eyes see. We can be blinded by such handy clichés. The notion of the "mind's eye," for example, propagates a widely held assumption that making something, especially when wrapped in the term "creative," is primarily a matter of executing an "image" held in the mind. Were that true, all drawing and making would consist of attempts at mimicking (an imagined) something, complete and immutable. Visualization and the even more misunderstood imagination undoubtedly exist, but the process is experienced somewhat differently: When we visualize something, we don't have or fix an image; rather, we arrive at motivation that accompanies an upwelling confidence.

This may be part of the preconscious brain preparing to act, but the point I wish to make is that visualization is not optical in the same way as seeing. Try to visualize a simple form, such as a cube. Close your eyes. What do you see?

Most people will claim to "see" *something*. I see nothing, but I *know* things such as faces, edges, points. I can even move and tumble the cube, but all that is done nonoptically. I cannot see it, but I *remember* it. I assemble it from simple pieces. For instance, I can't interrogate the surface for phenomenal features, such as reflections and shadows. The moment I go to draw the cube, everything changes—not unlike the way putting down words activates a sentence. I draw a line and the cube develops into view. Visualization happens in the moment, in the act on the page as much as in the mind (as a nonoptical memory). It is an act that brings past and future together in the present.

Returning to stone toolmaking—or working in any medium, for that matter—we need to consider the structure of visual memory. We visualize as much with our bodies as with the "mind's eye." The tacit knowledge embedded in our physical activities is, in some respects, where visualization can be found. Try drawing a human figure from life; after some practice, a drawing can be made from "visual memory"—a memory that amounts to muscle memory as much as eye and mind conceptualization. It is the ability to understand tacitly what an image is of.

In our brain's trifurcated memory storage system, *procedural* memory—how we know how to do things—stores generalities about learned procedures and sheds temporal recollections. If, for example, you remembered each "event" of a misthrown ball, it would be difficult to absorb a tacit understanding of the general procedure when throwing the ball. *Episodic* memory is what psychologists and neuroscientists call our ability to retain events in time and space; it does not deal with generalities, but rather remembers details and reconstructs specifics. Episodic memory is part of what makes us conscious and able to move mentally backward and forward in time. Finally, *semantic memory* stores ideas and meaning, often derived initially from episodic memory and then developed into concepts that replace the originating event(s). As Merlin Donald, a psychologist, neuroanthropologist, and cognitive neuroscientist, has explained, these types of memory are at odds with each other and occupy distinct areas in the brain.[20] In other words, you can remember *how* to do or *what* you did, or what it *means*, but one neural storage system cannot resolve all three types of memory.

While apes have a very substantial episodic and procedural memory, they cannot conceptualize in the way humans can. That is one reason they cannot draw or make the way humans can; they lack a semantic memory.[21] When apes are taught sign language, it is extremely challenging for them: they can access the sign's meaning only by remembering the context, the event memory of the sign—that is, by remembering what happened when they made a certain sign.

There is no doubt that early stone toolmakers had at least procedural and episodic capacity. Without semantic memory, toolmaking would have been much more instinctual and stable—as the archaeological record confirms. What is unclear is how semantic memory developed. Did it lead to or follow from toolmaking or image making?

Although the oldest-known art is predated by 2 million years of toolmaking, it coincides with the most advanced Acheulean phase of toolmaking. The so-called "Venus of Tan Tan" figurine, a work of our ancestral species *Homo erectus* that was found in Morocco in 1999 and is said to be the oldest sculpture ever unearthed, is only 300,000 to 500,000 years old. Since the advanced Acheulean toolmakers of that time employed much more planning and more dexterous procedures than earlier toolmakers, they would have had to rely on a robust semantic memory, the kind needed for image making—marking a significant milestone. Consequently, making could have shifted from a purely instinctual impulse to one with technical and design intent, from hard-wired species behavior to cultural action.

I am arguing that visualization, especially visual memory, brought the three types of memory into correspondence, liberating actions from the limits of the hardwired and heritably contingent, moving our ancestors to make and transmit artifacts, ideas, and beliefs based on intelligent choices. This marks the threshold of culture, which itself can be defined as the consistent intergenerational propagation of meaningful, nonheritable activity and thought.

What about the role of materials or media in this development, the shift from instinct to technical intent, from species behavior to design? The stones themselves may have something to tell us about this shift. The more primitive early Paleolithic tools were made from the cobblestones at hand, their simple fashioning attributed to the less-developed brain of *Homo habilis*. Acheulean toolmakers worked with flint, jasper, and other easily knapped stones. But when they did work with cobblestones, they reverted to the earlier primitive methods and tool forms, a clue perhaps to how making emerges from material as much as it is imposed upon material and how this influenced the evolution of toolmaking. This behavioral "relapse" refutes the notion that toolmakers were simply imposing form on raw material. Quite the opposite: the material was an inseparable component of thought and action; it guided the inquiry of what to do and how to do it. We hardly notice ourselves being influenced by materials in our contemporary culture, and probably would deny that it was happening. A charcoal-on-newsprint drawing brings forth attentiveness to light and shadow, phenomena readily captured by the medium, but the seamlessness of experience is such that we suppose our attentiveness comes from our volition. The age of machinery has long since turned us away from following the lead and guidance of material and toward believing that material exists to receive and submit to our intent.

The earlier *Homo habilis* were under less pressure to respond resourcefully to a changing environment. They were less prone to move and would by default tend to use available material. By contrast, the highly mobile *Homo erectus* migrated great distances, spreading out from Africa. A changing landscape, geology, and animal habitat must have challenged the mind, material now being a question of choice resolved only through experimentation. "Fooling around" may have had a role in the shift to more sophisticated toolmaking. When they took up using flint, its superior flaking properties may have been what ensconced the new techniques for the next million years, regardless of other changes—and would explain the reversion to old techniques when cobblestone was employed.

Stephen Jay Gould developed the idea of "punctuated equilibrium" in evolution, an important amendment to Darwin's theory, proposing that evolution is not a constant ongoing process but rather a mostly stable state punctuated by rapid change when selection pressures appear.[22] The phylogeny of toolmaking seems to follow this pattern. Acheulean toolmaking barely changed over the course of more than a million years and across an immense geographical area. Does this suggest that *Homo erectus's* brain power, intelligence, and volition were not yet independent of evolutionary "equilibrium," that a lack of selection pressure ensured that not only the species but its technological abilities remained in the firm grip of heritability and hardwired habit?

Species inertia may have kept this technology from evolving, but there are tantalizing hints of cultural development that would not have been possible without this monotonous million-year sameness. Some hand axes were very elaborately made—too elaborately for mechanical purposes—prompting speculation by archaeologists about a cultural and symbolic role for tools, for instance that they served to demonstrate some prowess, or were used as a medium of exchange. The sheer number of axes and excessive size of many is indicative of such a shift in purpose. This kind of cultural development could flourish only with the unchanging automatic technology of the toolmakers. Such a stasis—the technological equivalent of evolutionary equilibrium—indicates a (perceived) lack of functional pressure, allowing a technical artifact (such as a stone axe) to be transformed gradually to a representation of culture, first as a typical part of life and then as a symbol, a stable cultural element acting like language, complete with the encoding and syntax of speech.

This phenomenon persists in contemporary culture. We continually reconfigure artifacts no longer under functional pressure that as cultural objects derive much of their meaning and value from their design and aesthetic qualities, such as wristwatches and chairs; their utility is taken for granted and is rarely a differentiating factor. It is a recurring theme of this book: when functional obligations and pressures recede, the forces, action, and techniques organizing material are redirected to representation, to making things language-like. This tendency, which defines every culture, is perhaps the strongest evidence linking the act of making (manipulating material) with social and cultural order. Making things language-like can also be understood as making things mindlike, in that symbols and representations are what the mind operates with and on. This not only supports the enactive theory of mind, but suggests that the very act of enacting is transformative to both mind and material, conflating them into a whole.

On the surface, an impulse toward design and representation may seem trivial compared to advances in technology. It is not. Language forms culture—not only spoken language, but also the "language" of artifacts. As stone tools evolved from technological to cultural artifacts, so did the ability to organize great numbers of people, which in many respects exerted much greater leverage than technology alone in radically extending our ancestors' capabilities beyond their zoological limits.[23]

SOCIAL ORGANIZATION AND ART

It is easy to assume that the development of early art was a matter of progression from the simple to the complex. For example, a natural form may first have been modified slightly to represent a figure. Then, with increasing visual ability, more robust modifications of a natural surface outline were made to resemble an animal. This led, finally, to the wholly *de novo* work. But we don't really know, since the archaeological record provides so little, especially at the boundaries of paleontological and cultural frameworks of archaeological inquiry when art making appeared.

What were the early motivations for art making? If we go back to hunters and their short, brutal lives, we might suppose they were preoccupied with survival and procreation, under constant stress from the unpredictable world in which they lived. The shortness of life precluded anything like the long period we now devote to education. What could get passed on was mimetic, training by practice—another possible explanation for the consistency of early stone tools. Then bands grew into something larger.

As with any group, the larger they became, the more complex their organization. They had to develop a division of labor, rules, taboos, power structures, and all sorts of intricate hierarchies and processes. Complex social organizations required a worldview, the construction of a mental model to bestow stability and survival benefits on the group.

The model itself was key to developing a society's ideas. Unbound by an unchanging mimetic reproduction of techniques and tools, a mental model—albeit one that evolves slowly at first—could adjust to pressures more quickly than in biologically dependent (evolutionary) change. Rites, norms, games, images, technology were all agents of the model, and as such became the structure of society. Social structures operate collectively: everyone speaks a certain way, operates within the same rules, plays the same games, and practices common rites. Image making was unique in that it required a skill set beyond heritable traits, skills that only very few especially talented individuals could have possessed (as remains the case today).

Despite this limitation, image making was pervasive throughout Europe and Asia from about 40,000 years ago, which coincided with the arrival of *Homo sapiens*. That, in turn, suggests an even earlier origin. As dates for image making keep getting pushed back, it is possible to consider the beginnings of language and image making as contemporaneous.

Both are representational schema, marking a crucial cognitive boundary between us and other species.

Paleolithic images from far-flung locations, made over thousands of years, are remarkably consistent in manner and subject matter and almost always found in inhospitable, difficult-to-reach, dark depths. This consistency implies cave art was part of a social structure, and most likely of religious significance. As in today's religious organizations, the persistence of traditions and artifacts over great time lends credence and even verity to that religion—a positive feedback loop encouraging steady and unchanging production of its images and rituals.

Cultures, as shared systems of symbolically communicated knowledge, are built on societal structures or models. The earliest cultures were mythic, built on narrative to externalize and unify the world "model" represented by image making, artifacts, and language. Their transmission across the generations depended on conventional representations, including images—a description that applies to Paleolithic cave art. Early artists were more like artist-priests charged with the responsibility to maintain and propagate the narrative structures of their culture and, judging from our own inclinations, to sense a connection between actions and unrelated events. The magical power of images must have been deeply felt, a religious experience. The ecstasy of miraculously transforming the utterly sensory-depleted, empty darkness of the cave into a cockpit facing an illuminated window onto the world—the world as representation—conjures so many of the most basic religious archetypes that it is difficult not to draw a line directly from cave art to religion. And despite the *individual* creativity of these artists—or because of it—their art was an expression of a *social* imagination or, in Jungian terms, of the *collective* (un)conscious.[24]

Consciousness did not arrive as the tree of reason nor the wind of the imagination blowing through it, but as a strange, low, rising howl as our evolutionary progress reached storm strength. The transition from evolutionary (Darwinian) pressures to cultural ones as the chief driver of heritability happened when our representation of the world became operational—when we began living in the story we told ourselves and each other, when our actions began filtering experiences through the new eyes of a worldview constructed by the self-conscious mind.

In his essay "A Defense of Poetry," the English romantic poet Percy Bysshe Shelley (1792–1822) wrote, "Reason is the enumeration of qualities already known; imagination is the perception of the value of those qualities, both separately and as a whole. Reason respects the differences and imagination the similitudes of things."[25] Imagination is the echo of the former unity, forever sensing and searching for that lost innocence. In terms of the hunter's worldview, the relation to prey became so much more complex; it took imagination to recreate the oneness with the prey, in the form of empathy, a mental construction of similitude. Reason was necessary to evaluate conditions, to measure progress and distance and stamina, but imagination was needed to place the hunter into the future, beyond the "now" of the hunted animal.

3 WORDS MATTER

Words and language may seem unlikely subjects in a discussion about materiality and materials. The mind, it may be argued, is the exclusive purveyor of words—they don't exist elsewhere other than as representations. But when we began to live in the stories we told, and when actions began filtering experiences through a worldview constructed by the self-conscious mind, language and words could not be separated from their material origins, bringing forward the possibility that material is far more embedded in and influential on the mind than we might realize.

Words are immaterial, and we are hard pressed to call the act of speaking and reading "active engagement" in any strongly physiological or material sense. I can speak to a stone, but I doubt it will be affected by my words. Language, however, cuts as sharply as a knife when directed toward other humans. Language in all its forms—speaking, writing, reading—is also embodied engagement, not with material but with others, functioning as the prime agent of social organization. This doesn't remove language from involvement with material; these next chapters explore just how language has come to influence material and to be influenced by it, even to the point of affecting thought.

If we observe the effects of language at the greatest scale—from its inception to the present, at a global level—we can understand its power. André Leroi-Gourhan observed that throughout our species' history it has been our ability to organize ourselves into complex functional groups, far more than our technological advancements, that has leveraged our zoological abilities.[1]

Language has turned humankind into a gigantic social organism, or at least a collection of them arranged by language, making it possible to form elaborate networks and social orders that, in turn, have resulted in tremendous changes to the physical environment. Instead of being ineffectual in dealing with material, like shouting at a stone, language turns out to be highly effectual, the agent organizing groups of individuals to hammer, chop, lever, roll, pull, wedge, and otherwise cooperatively break the resistance of the material through a complex multitude of steps. Language extends and embodies action from the individual to the group to larger and larger social sets; it is this "collective embodied action" that meets material, transforms it, and, in my opinion, is also influenced by the behavior of material.

But how? Material conditions directly affect social organization; how we organize ourselves to extract and process material or make things is contingent on physical properties. Agricultural societies have always organized themselves based on the material conditions of the environments in which they are working. Cost and price encode the collective efforts involved. But as the chain of connections between individuals and material grows longer, apperception of that relation fades. Who can grasp the physicality of the particular species of tree that lies within a piece of paper? Language, though, can transmit more than the individual senses can receive. Words carry material characteristics back to us as little coded encapsulations that, upon being received, activate sensation, even if our senses have no direct access to those characteristics.

Two words, "artificial" and "natural," are a simple example. We call plastic "artificial" and paper "natural" not because plastic is not made from natural material, but because its manufacturing process involves steps that are outside our physiognomically understood universe. Even without knowledge of the technology involved in papermaking, we understand and feel that a tree can be transformed into paper, and hence the word "natural" comes easily to mind. The word "artificial" applied to plastic is, in great part, the transmission of a cultural idea that originated from a much longer chain of collective embodied action transforming raw material.

Language, as the primary medium of thinking, travels between the individual mind, the larger social order, and the physical world, seeming to bypass the body. If we look a little deeper, however, a more embodied and materially influenced structure of language emerges. This is made clearer by examining how our physical surroundings and physiognomy have been involved in the origins of speech and of the process of thinking and in the development of writing.

FROM GESTURE TO THINKING

Of all the mysteries of human evolution, the origins of language are among the least understood. Since language was oral long before it was written, we have no record that would allow us to see a gradual transition from gesturing and mimicking (including *onomatopoeia*) to the use of symbolic language. For the most part, we can only speculate from the so-called "writing" on cave walls and elsewhere.

The gap between gesture and language is daunting, since what is required is the ability to abstractly signify common elements. According to V. S. Ramachandran and Edward Hubbard, researchers in behavioral neurology and psychology, the brain's architecture offers some evidence for how seemingly unrelated things such as an image and a sound could be brought together in a meaningful and systematic way.[2] The brain's processing of language is modular, evidence of language evolving from other functions. This isn't unusual in evolution; the wing's origin was as a cooling appendage, and only later shifted its function to flight.

Language involves extensive networking among many areas of the brain, including those of motor control of the mouth, lips, and larynx; syntax control; and semantic and auditory processing. These centers in the brain may have evolved from physical and material conditions; Ramachandran, for instance, speculates that Broca's area, the part of our brain that controls syntax, initially evolved from the increasingly complex procedures of toolmaking and was only later leveraged by the drive to language.[3]

Still, the question remains: How did a gesture or a sound become abstracted and related to something that wasn't a gesture or sound? A compelling bit of evidence can be found in the relationship between images and words. German psychologist and phenomenologist Wolfgang Köhler

(1887–1967) in 1929, and Ramachandran and Hubbard seven decades later, tested subjects by asking them to associate two images and two arbitrary (meaningless) sounds.[4] Ramachandran and Hubbard discovered a high degree of correlation (95%) between an image of a rounded shape and the sound *bouba* and between an image of a jagged shape and the sound *kiki*—what is now known in the psychology and neuroscience literatures as the Bouba-Kiki Effect. How could this be?

One explanation put forward by Ramachandran and Hubbard is that the shapes our mouths make when saying *kiki* and *bouba* emulate the respective images. Language is full of other examples: *mush*, *crack*, and *fudge*, for instance, require the mouth to mime the material characteristic of each word. It is the best evidence of "cross mapping," in which an image is mapped to a sound that is related to a motor function (shaping the mouth)—a basic networking of independent modules laying the groundwork for the abstraction necessary to create words as a mirror of reality.

While abstraction is the key to language, it should be noted that even with the Bouba-Kiki Effect, materiality should not be overlooked. A jagged shape associated with the *kiki* sound is first and foremost a fact of matter. Breaking a stone results in a "cracking" sound and a sharp, jagged edge. Breaking a piece of glass has the same visual-sound correspondence. All such jagged edges are tied to sharp *kiki* sounds as part of lived experience. Hence, it is possible that material manipulation (making tools, shelter, clothing) led to a heightened ability to establish cross mappings among materials' visual, auditory, and tactile properties. The unified nature of the source material would only reinforce the necessity of neural networking across the senses and those neural modules arising from material engagement.

How much of a role did primitive technology and the systematic manipulation of materials play in the development of language? Material's concreteness and language's abstraction would suggest that they have little to do with each other, but as the Bouba-Kiki Effect shows, materiality seems to reside deeply in the structures of language.[5]

The influential "recapitulation theory" of German biologist Ernst Haeckel (1834–1919)—that "ontogeny recapitulates phylogeny"—proposed that an organism's embryonic development (ontogeny) followed a course that corresponded to that species' evolution (phylogeny).[6] While no longer widely held in its original field, Haeckel's theory has continued to be used in other fields of inquiry, including studies of the origin of language.[7] One question it suggests is whether a child's acquisition of language might offer insights into the evolutionary or historic origins of language.

Children acquire language in synch with their physical development—movement accompanies words. According to German psychologist Wilhelm Stern (1871–1938), perhaps best known for inventing the concept of the intelligence quotient (later used to develop the first IQ tests), the first word a child speaks is accompanied by a gesture and body articulation—sound, gesture, and "embodiment" are articulating more than a word.[8] Accompanied by a hand motion and a body curl, for example, "Mama" uttered by the child is actually "Mama, give me that."

Soviet developmental psychologist Lev Vygotsky (1896–1934) observed that the word (in this case, "Mama") was deployed as a substitute, an oral reinforcement for the gesture.[9] The gesture was still the key communicative element—the exact inverse of our adult conventions of speech and gesture, in which gesture plays only a supporting role. All this happens long before there is any conceptual grasping and manipulation of language. If we consider the evolutionary question, this moment in a child's development easily overlaps that of our primate relatives.

Volition, *action*, and *touch* must work in unison for reality to become real—that is, to become perceived and categorized. And so it is with the young child. Once capable of combining volition, action, and touch, she becomes mobile and articulates her hands, naming becomes prolific, and the first of two fundamental attributes of language comes into being: the arbitrary encoding of things with sounds (words). It is arbitrary in the sense that the words operate symbolically, not onomatopoeically or as transitional sound symbols (such as *bouba* and *kiki*). This naming, this reaching out and "touching" the world with the *what* of curiosity, is *touch*, running the hands over the world to feel it, to know it. Through repeated action, a mental map—naming and ordering things—is slowly built, each iteration altering the previous one as experience invokes new selections and, in turn, new experiences.

As children develop, they begin to assemble small sets of words, linking action to object/subject. This level of language has been achieved by some of our gifted primate cousins, but this marks the limit of language for them. What follows for the child is what makes us distinct from all other animals. While there's plenty of discussion about whether language ability is hardwired, the child continues to acquire language—which now propels her socialization, which in turn propels language acquisition. Children begin to construct increasingly complex sentences and to work with linguistic rules and abstract concepts. Curiosity about the *what* of the world moves to the *where, how, when,* and ultimately to meaning—the *why* of things. Language becomes thoughtful, becomes thought.

This transformation to "human" is somewhat of a blur. How did it happen exactly? How did the child acquire an "I" that can imagine the future, that can view the world conceptually and explain things (through narrative)?

Vygotsky proposed that language and thought are different and independent developments in the human brain that come together just as the child learns to speak, or, more pointedly, that thought "hijacked" language—giving thought a vehicle or means to be self-aware.[10] From Vygotsky's observations, there are three stages of speech in children. The first, social speech, is used to communicate with others; it typically develops at age two. Private speech, the second stage, is transient in that it appears for a limited time in life, typically from ages three to ten, after which it disappears. Private speech, children talking to themselves, is not for communication; rather, the child is regulating herself—talking herself through a situation. Vygotsky observed that private speech occurs predominantly when the child is challenged, speech being employed to organize the self and plan actions. Speaking is thinking out loud.

Other psychologists[11] have confirmed Vygotsky's observation that children using private speech are using it as a tool to leverage cognitive abilities, to imagine, to describe to themselves what is happening—as if they cannot yet "think" what is happening. Thought development is dependent on entering oneself from outside: *I need to hear myself think*. According to Vygotsky, private speech diminishes in audibility between ages seven to ten or so, but it doesn't actually disappear. Rather, it goes underground, transforming into the third stage, inner speech (including the self-regulating function of our so-called "internal monologue"), our familiar nonspoken yet word-formed thinking process—*verbal thought*.

Private speech is tied to social development. The more socialization children experience, the sooner and more extensive the use of private speech. The implication is enormous in terms of phylogeny. Thought and language are products of socialization, and thought's coupling with language is linked to an extrinsic factor—the social group. Is it unreasonable to conclude that a simple increase in group size of early hominids (the result of effective toolmaking?) put selection pressure on the brain to increase its capacity to handle the increasing complexities of large social organizations and, in doing so, coupled language to thought, thus bringing consciousness into being?

LANGUAGE AND EMPATHY

Language allows us to interact and communicate socially in complex ways, but its relation to our more unconscious or instinctive traits is less clear. Empathy appears as a trait separate from but affected by language. How we can understand and feel another person's frame of mind seems an attribute of our neurological architecture, not the product of language. This may be partly true (social animals are empathetic). However, because of the way language is acquired—through much social interaction—the very act of sharing experience coded in language creates a path for empathy: I can feel for you when I hear you tell me your story; I can weep when I read about someone's suffering.

Mirror neurons in the brain—which literally reflect as one's own experience what is perceived in others—are among the many structures that independently and in concert make us empathetic creatures. We have what is called *cognitive empathy*, the capacity to understand the perspective or mental state of someone else. We also have an automatic form of empathy referred to as *affective* or *emotional empathy*—our capacity to respond to another's mental state with an appropriate emotion: our physiology and emotions react to others without the use of language or conscious thought. Does this capacity for empathy extend to our inanimate physical surroundings as well? Does our feeling for things, including materials, stem in part from our empathic ability? Does language, in providing a path for empathy, also play a role in our empathizing with materials? Anyone working with material, including craftsmen, artisans, or the car clay modeler mentioned earlier, would agree that material empathy is real; for them it provides the agency to feel into and follow material.

THE MATERIAL ORIGINS OF WRITING

The development of writing is a story of the meeting of language and material in which the concept of written language was enacted from the particular traits of a material. Writing unfolded from slow and incremental steps, typical of most human discovery, providing an excellent example of just how richly the mind and material can extend and enact one another. The story of writing involves one material in particular, clay, in soft, hard, and fired form. Had some other material been in use in that society, writing would probably not have been invented, or not in the way it was. As I mentioned earlier, materials resist, and in their resistance to our actions they stimulate the imagination and creativity. Materials also lend themselves to certain potential actions that can, as we will see, play an enactive role for the mind.

The American psychologist James J. Gibson (1904–1979) coined the term "affordances" to describe the potential an environment might hold for an animal or human.[12] A common example is a handle on a teacup, an affordance for grasping, or a stair, an affordance for climbing, but only to species that can climb those stairs. The debate about affordances and where they are located—whether they are learned in mind or latent in the environment or some ghostly in-between—is suggestive of the assimilations of material and mind.

One might further consider *material affordances* as being the affordances a raw material might offer, in the sense of the potential it holds. Paper has the material affordances of folding, crumpling, and tearing. Clay has the material affordances of being easily formed (plasticity), fired (converting from soft to hard), marked, impressed, and imprinted. Both the resistances and affordances of clay guided the development of the abstract conception of numbers that led directly to the development of writing.

We are quick to separate physical from mental actions and have confined each to separate institutional realms. Legislators don't draft laws by whittling wood, and artists don't paint landscapes by reading. It's surprising, then, to consider that writing most likely evolved from making. The story begins with thousands of mysterious little clay and ceramic forms found scattered in Neolithic archaeological sites in Mesopotamia. When they were first discovered, no one had any idea what they were. Some speculated they might be game pieces.[13] Archaeologist Denise Schmandt-Besserat pieced together how these little objects came to enact greater and greater degrees of abstract thinking and ultimately writing, undercutting the long-held assumption that writing had evolved from coded images (pictography). The intriguing story behind these "tokens"—some of the oldest known examples of fired clay and ceramics—is one of how ideas and thinking are born and borne through material and how material's "concreteness" can be critical in propelling abstract concepts.

As with all design processes, the historical development of counting was *incremental*, each step leading to another; *accidental*, an insubstantial feature from an earlier phase becoming a critical element of a subsequent period; *static*, only evolving when under functional pressure (when the existing system wasn't working well); and *arbitrary*, the significance of form and figure being assigned rather than inherent. Of all these conceptual underlays, arbitrariness may have been the key, allowing an understanding from the start that *this*—*this* being something that had to be learned—could stand for *that*, and by implication *that* could be exchanged for a different *this*.

We take for granted the concept of numerals, but only because we teach it to our children; number sense—at least beyond about the number three—is not innate. The earliest record of counting is found in "tally" sticks dating back 20,000 to 30,000 years. Marks incised on bones denoted something, but exactly what isn't clear. Some archaeologists propose they refer to the lunar cycle or other kinds of time counting.[14] In that prehistoric period, societies were generally egalitarian; hunter/gatherers lived from day to day, nomadically, in an opportunistic economy, obviating any need for accounting. Why bother when things are shared?

The advent of agriculture and settlement created a new type of economy based on surplus, storage, and commodities. New social organizations developed in parallel, with the emergence of hierarchy of position or rank and the division of labor. The need to keep track of things and people's efforts and property increased exponentially. Tally sticks, or piles of gathered twigs and pebbles, were used early on and had no meaning other than to record. Then, about 8,000 years ago, a new conceptual device appeared. Analogous to the shift from the ad hoc nature of hunting/gathering to agriculture, gathered twigs and pebbles were supplanted by manufactured ceramic tokens that brought a new level of systematic organization to the task of keeping track. Conceptually, these tokens were a form of "concrete counting," not related to abstract numbers. Most novel was the shift from generic marks or anonymous pebbles to the assignment of a specific meaning to each ceramic form. For example, a small cone meant a specific measure of grain; a simple ovoid represented a jar of oil. In figure 3.1, the cone, sphere, and flat disk are three measures of cereals (small, larger, largest) and the tetrahedron is a unit of work (perhaps one man/one day). The 5,000-year journey on the road to writing had begun.

Clay acted not as the medium to receive a formed idea but as the enacting agent. The tokens were simple, easily formed by hand from soft clay. The resulting cones, cylinders, discs, ovoids, spheres, and so on also shared the characteristic of being cognitively simple, making it easy to convey the

3.1
Tokens from Tepe Gawra (present-day Iraq), c. 4000 BCE.

technology of making them to others—even with spoken language exclusively, in lieu of any physical demonstration. I can describe a cone to you over the phone and be reasonably certain of the outcome. The firing of clay, of course, turned these soft artifacts into "stone," achieving the durability of pebble counters but with the new encoded meaning. Clay's enactive role, though, did not end there. Each stage in the evolution of the token system as it evolved into writing was generated with and by the affordances and resistances of clay/ceramics.

The first simple tokens were easily formed by hand, but the limits of such a class of forms were soon reached. As more commodities were added to the token system, the tokens began to change and become more complex (figure 3.2).

As one might expect, tokens took on more elaborate forms, but they also began to incorporate marks, lines, dots, and x's to overcome their formal limitations. A similar form could represent different commodities by

3.2
Complex tokens from Uruk, Iraq, c. 3300 BCE.

being marked with a different set of lines. Here another incremental and necessary component of writing and abstract counting was falling into place, namely the idea of making a distinctive mark to mean a distinctive quantity and type of thing (but still tied to form). The material, clay, plays a crucial role in such articulation. Because of clay's specific capacity to be shaped easily into simple forms and to be tooled (marked with an instrument), this development of the complex tokens cannot be conceptually disassociated from the material. The concept was derived from the material traits.

The token system was successful and worked well enough to persist for thousands of years. Eventually though, shortcomings—brought on by the rising complexity of organized society—overloaded the system, preparing the ground for change. The first new step was the introduction of "envelopes," which consisted of hollowed clay balls used to store tokens inside. In figure 3.3, the lenticular disks each stand for "a flock" (10 animals, perhaps); the cones are measures of large and small amounts of grain.

3.3
Envelope from Susa, Iran, with tokens to be stored inside, c. 3300 BCE.

Here again, the plasticity of the clay helped generate a new concept—that of holding records in sets. Clay ball envelopes were the distant antecedents of file folders, and were likely used for recording debts and contractual obligations. The tokens were placed inside the soft clay envelopes, which were then sealed, producing an archival record. In solving one problem, however, the envelopes created another, which proved to be the catalyst for an even more significant conceptual advance. The immediate problem was a lack of visibility of the contents once the envelope was sealed. The solution was to impress the tokens into the exterior of the still soft clay of the envelope before they were placed inside it. The impressions were easy to understand, since they were a physical record like a footprint, but they also opened a conceptual space wherein an impression or mark was understood to represent something that did not need to be present. The freestanding mark was born. Concreteness moved slowly toward abstraction.

Imagine methodically pressing little cones, spheres, and cylinders onto the exterior surface of a soft clay ball, then placing them into the hollow and sealing it. On the outside of the envelope, the pressed marks are the only visible record of the contents. The way the tokens are pressed is key; a cylinder pressed with its flat base could be confused with a disc. Each token, then, must be pressed in such a way as to maintain the *legibility* of its geometric form.

Another problem was that the tokens were never intended to make clear reliefs, so the "scribes" had to be careful not to press tokens at an odd angle; styluses were developed to make some of the requisite reliefs. However they were made, these impressions were no longer fully three-dimensional; halfway to two dimensions, they portended a new concept—reading.

Even with the advent of marks on envelopes, counting was still concrete. Each impression stood for a unit quantity of a commodity. The idea of a disassociated number still lay over the horizon.

According to Denise Schmandt-Besserat's research, the next stage of the Mesopotamian accounting system's evolution saw the envelopes superseded by tablets once it became evident that the contents of the envelope were no longer needed, just their record. This wasn't done in a single leap. Early tablet versions were convex, echoing the convex surface of the spherical envelopes, only later adopting a flat format. Here again, clay's intrinsic materiality both generated and limited the changes in the accounting system. The tablets continued to be made with soft clay impressed with tokens, the meaning of each form remaining stable. A cone, for instance, still stood for a measure of grain. Now, however, the *impression* of the cone was authoritative, so much so that the actual cone was no longer needed. The mark superseded the form.

As Mesopotamian society became more complex and technologically advanced, the increasing variety of products and commodities led to more elaborate tokens. Arbitrariness (of form) was an important attribute of the token system; this allowed for the creation of more and more token types. The vocabulary of forms was also expanded with systematic incised marks. Two tokens could share a form, but each, because of its unique markings, could refer to a different commodity. This paralleled the marking of the envelopes, furthering the conceptual shift from exclusively form-based reckoning to a hybrid of form cognition and mark reading.

It is interesting to note that envelopes were all found to contain the more basic, unmarked tokens (which still referred to foodstuffs). The reasons are not clear: it may be that envelopes were used only for agricultural commodities. Regardless, it does point to a problem with marked tokens: while simple forms can be recorded by a physical impression, the marks on tokens, being shallow, do not lend themselves to being transmitted by impression. What to do? The clay's properties were a heuristic. A range of approaches, most becoming dead ends, have been documented, including pressing the tokens directly into the exterior surface (no translation needed), scratching marks into the hardened clay envelope (very difficult to control), and, of great significance, adding tooled marks to the impressions of the tokens on the outside of the still soft envelopes. Disassociation, which had eluded concrete counting, now appeared in the actions of recording. Token counters were used to make impressions, and scribing tools were used to add marks. Qualities and quantities were no longer cognitively locked into a single form-concept.

Tablets carried on the combination of impressing with tokens and marking to carry all the information required. As tokens became more elaborate in terms of their incisions, it would have been evident to the scribes with their styluses that the markings on the tokens were more readily "written" onto the tablet than by impressing the token forms, setting the stage for the final step toward writing.

FROM SOUNDS TO MARKS

Imagine scribes busily pressing tokens into clay tablets and then adding the increasingly elaborate incised markings of each token type. It is clear to all that the tablet is the artifact of record and that the token is no longer the referent. The work isn't easy, a lot of tokens need to be at hand, the placement of the tokens and marks is strictly prescribed, and each month the work days seem to get longer and longer, as more production is demanded from the small corps of trained scribes. But no one is willing to challenge the procedure. "This is how we've done it for 300 years. Why fix what isn't broken?"

One day, a young scribe is rushing to get through the backlog and decides to break down his workflow by taking a set of ten tablets and, using his stylus, outlining the placement and type of impression he will make with the tokens. His assembly-line approach speeds the work until a supervisor happens along just as the young scribe has finished premarking a set of tablets.

How does the story end? We don't know, but we do know that pictographic representations of tokens made their appearance after nearly 4,000 years of continuous token use. The final disassociation between quantity and type was achieved by using pictographs to register the type and impressions to register quantity—for example, a tablet from 3100 BCE has an inscribed picture of an incised ovoid figure (whose antecedent was an incised ovoid token that stood for a jar of oil), while next to it are impressed marks: three wedges, each standing for 1, and three shallow circles/discs, each standing for 10 (figure 3.4). The tablet is describing 33 jars of oil. Numbers had finally become abstract, disassociated from anything physical.

Once numbers were in place, the development of writing relied less and less on the traits of soft or hard clay. Pictographs continued to be

3.4
Tablet from Godin Tepe, Iran, c. 3100 BCE.

marked in clay, as were later cuneiform types of writing (so named because of the wedgelike markings made by styluses, *cuneus* being the Latin word for "wedge"), but the real action took place away from the limits of the clay.

The desire to have the living speak one's name after death, to carry the spirit into the afterlife, may have pushed spoken language to aspire to the kind of permanence of data preservation seen in accounting.[15] Whatever the reason, tablets began to be used to record names, but since names had no shapes but only sounds, phonograms—pictures that corresponded to sounds—were invented. We sometimes find contemporary examples in the children's section of the Sunday comics, where images are added together to make up a phrase—a so-called "rebus." Speaking from a visual code had arrived, the last major conceptual piece leading to what today we call writing.

There were, of course, many further developments that brought written language to our time, but never again were developments so contingent on, generated by, and enacted through material. In this case, clay had acted as a rocket launcher, sending the project of counting and writing on its way.

Once writing had become the cerebral act we know it as, material played only a supporting role in its further evolution. It did not disappear completely from writing, only as an agent of writing's development. Material reappeared in vocabulary and within the very structure of thinking. Metaphors, the basis of so much language, became building blocks of thought (yes, that itself is a metaphor), lending concrete sensibility to abstract concepts—an echo of the story of writing in which abstract ideas emerged from concrete sensibility.

4 REPRESENTING

I have discussed the role of materials in the origin of both spoken and written language and how language is, in essence, a medium of the mind. In this chapter, I probe how representation mediates our relation with the surrounding world, providing an essential means of manipulating physical reality, and is in turn deeply and mostly unconsciously influenced by our embodied experience of that same physical reality.

I limit the term "representation" to how we relate to the world around us through the transformation of apprehended reality into a symbolic construct, which defines consciousness. The means and media of representation are not limited to language: images, dance, artifacts, and physical phenomena are persistently overlaid with cultural signs and symbols and understood in that light. It can even be argued that the domain of our awareness and attention is limited to what we can represent (and thus communicate), ruling out perceptions of reality that lie below the threshold of consciousness.

Representation is paradoxical in that it limits our conscious access to reality while liberating us from physical constraints, giving rise to the possible, the imagined, the not yet determined.

Of particular interest here is how visual/spatial/material experience can correspond to language to become symbolic and communicative, and how the material terms of physical reality have come to reside within the abstract terms of language.

How embodied experience informs representation eludes complete understanding because of the inaccessibility of the unconscious processes that link the sensory to the mind. Outside the mind, in the structure and nature of language and images, we can find better evidence of how the experience of physical reality and the mind's constant reordering of a symbolic universe intersect and transpose onto one another. Ezra Pound's famous two-line poem "In a Station of the Metro" offers a precise example.

The poem is structured as two "images"—one from experience and another "of the mind"—that meet, each altering the sense of the other. The poet uses language to create images, meaning in this regard that he attempts to bring language closer to the unconscious dimension of sensation by stripping away any symbolic (representational) content. Hence, the movement in early twentieth-century Anglo-American poetry this approach sparked was called "imagism."

> The apparition of these faces in the crowd;
> Petals on a wet, black bough.[1]

Yes, writing can be more powerful than sight itself. Words here make an image that is more richly associative and "real" than almost any picture. A dark train moving by, pale faces momentarily lit, souls on a subway, life's gray, rainy ambience, a frozen moment before it all vanishes. The imagination opens, triggering visual and emotional associations.

Pound described the poem as having been stripped down to a pure experience, without rhetorical devices, flourishes, or attempts to represent feelings or interpretations. The power of the poem's language came from a long period of writing. His description of how the poem emerged is worth quoting at length.

> Three years ago in Paris I got out of a "metro" train at La Concorde, and saw suddenly a beautiful face, and then another and another, and then a beautiful child's face, and then another beautiful woman, and I tried all that day to find words for what this had meant to me, and I could not find any words that seemed to me worthy, or as lovely as that sudden emotion. And that evening, as I went home along the Rue Raynouard, I was still trying, and I found, suddenly, the expression. I do not mean that I found words, but there came an equation . . . not in speech, but in little splotches of colour. It was just that—a "pattern," or hardly a pattern, if by "pattern" you mean something

with a "repeat" in it. But it was a word, the beginning, for me, of a language in colour. I do not mean that I was unfamiliar with the kindergarten stories about colours being like tones in music. I think that sort of thing is nonsense. If you try to make notes permanently correspond with particular colours, it is like tying narrow meanings to symbols.

That evening, in the Rue Raynouard, I realized quite vividly that if I were a painter, or if I had, often, that kind of emotion, or even if I had the energy to get paints and brushes and keep at it, I might found a new school of painting, of "non-representative" painting, a painting that would speak only by arrangements in colour. . . . The "one image poem" is a form of super-position, that is to say, it is one idea set on top of another. I found it useful in getting out of the impasse in which I had been left by my metro emotion. I wrote a thirty-line poem and destroyed it because it was what we call a work of "second intensity." Six months later I made a poem half that length; a year later I made the following hokku-like sentence ["In a Station of the Metro"].[2]

Pound worked subtractively, expending his energy on removing words in order to rediscover the first impression he had in the subway—not a memory of it, but the original *emotion* that had come to him not as words but as color: The poem's doubled images, without recourse to tropes of representation, manage to evoke strong "emotions" without a single word conveying any emotion or feeling. The mapping of the poem from objective observation ("The apparition of these faces in the crowd") to inward imagination ("Petals on a wet, black bough") synthesizes perception's mix of visual concepts and visual phenomena.

Pound managed to use words to make something that can be described only as an intense image—the intensity of which is pure feeling.

EVERY STORY TELLS A PICTURE

Language and image making interweave, which suggests a common origin. The exact origins of spoken language and its coevolution with image making, however, are lost to us; the only record is the interconnected visuo-motor and language neuronal structures in our brain. When we draw, language is present; when we use words, we make images.

The famous aphorism attributed to French poet Paul Valéry—"to see is to forget the name of the things one sees"[3] ("Voir, c'est oublier le nom des choses que l'on voit")—describes the interdependence of visual- and language-based consciousness. A common exercise for students is to draw a stool upright and then draw a stool upside down. Invariably, the upside-down drawing yields a more carefully observed and sensitive result than the overly familiar and linguistically coded right-side-up "stool." The word and image of "stool" reinforce each other, making observation with fresh eyes difficult.

Another exercise, one I have given to my beginning students, asks them to make the form of an object to which only their hands have access (inside a box with two sleeve-like openings). At first, there is a lot of groping about and reliance on touch, a sense of blindness pervading the room. The very awkwardness of this way of working is indicative of how much the visual sense dominates. As the students begin to sculpt, they instinctively project and overlay a conceptual schema, which dominates the actual and felt attributes. Decisions are made such as aligning edges and making planes coplanar despite what the hands measure. It is clear that the mind-eye axis has overruled the eye-hand axis.

Only when the object remains unfamiliar or unnamable do the students trust their sense of touch to act instrumentally. Without the hand, without direct engagement of physical materials by the senses, the mind-eye relation remains bound to language (including the language of geometry) and its preconceptions.

Research indicates that a part of language memory is stored in quasi-image form.[4] For example, "cup" is remembered as a two-dimensional cartoonlike fusion of image/word, resistant to separation like the upright stool from its preconceived image/word in the student drawing exercise. If I try to envision some event from a while back, a vague "image" might form; but if I recall what was said at the time, the scene sharpens and fills with sensation. In this instance, the spoken words spark more associations across all sorts of experience memories than the compact recall-ready image/word memory.

Cognitive linguist George Lakoff and philosopher Mark Johnson describe such image/words as "basic-level categories,"[5] which define most of our knowledge. In their view, a basic-level category such as "cup" describes how human cognition orders the world. If we are too precise (e.g., "Meissen teacup") or too vague (e.g., "tableware"), we cannot form an image that clearly distinguishes the category. Lakoff and Johnson argue that images of such basic-level categories are always tied to how we interact with them physically. That defines the category as well. This is key to the authors' proposition that our language, categories, concepts, and even our philosophical reasoning are not independent constructions of the mind, but are instead inextricably linked to our embodied experience of material, spatial, and physical reality. This linkage, they say, can also be found in the way all concepts expressed in language evolve from the bodily experience of concrete and physical sources.

Language breaks down the world into bits and recombines these bits into new relationships. When we name, we separate things from experienced phenomena. Drawing images of things, because it is an act of naming, also separates those things from their continuum. When children first acquire language, they name things they directly experience or feel, thus encoding the physical world in abstract representations. The abstraction of language allows the mind to conceive, to understand relations between things, to articulate interior mental states, to express concepts such as "know" and "remember" and "might be"—thus allowing us to organize, structure, and categorize the world, to represent it as a reality, overlaying the incarnated world of sensory experience.

At its core, human consciousness is a meditation on concrete and abstract realities that are far more intertwined than we are aware. Much like matter and antimatter, these two—sensation and thought—are strangers, ignoring each other's lack of awareness. Recall how often "thoughtless" and "senseless" have been used as a critique of the other. Casually mixing abstract and concrete terms to describe an experience—consider, for example, "thoughtful storm"—often fails to evoke what was intended. By contrast, successful representations concretize the abstract, bridging from one form of reality to another, generally traveling from the shared world of physical sensation to the inner world of ideas, thoughts, and emotions. The meaning of *meaning*—which is the measure of the subjective—lets us understand why physical experiences are our mental yardsticks; these experiences allow us to apply concrete terms to abstract things to measure them—to make them meaningful.

METAPHOR

If language is the "language" of the mind, what is the "language" of physical reality? Phenomenologists employ "bracketing" as a tool to remove language from playing a role in direct experience; it consists of putting aside the terms used to describe something, obliging one to look deeper and deeper into the raw experience without the handy scaffolding of language.[6] A metaphor (from the Greek *meta*, meaning "over" or "across," and *pherein*, to "carry" or "bear") is a bridge between the concrete world and the abstract mind, linking something known to the senses with an abstraction of the mind, always working to describe the abstract with the concrete: "Conscience is a man's compass," as Vincent van Gogh wrote in a letter to his brother Theo.[7]

Researchers have measured brain response to metaphors; the greater the distance bridged, the greater the pleasure.[8] This tells us of our hunger to find associations between the senses and the mind. The social and cultural provenance of metaphor is so distinct that we barely notice that metaphors are strange and arbitrary, yet instantly understood. He is "the apple of my eye" cannot be modified to "the orange of my eye." The former is understood culturally and metaphorically, the latter literally.

A good metaphor can feel more "real" and authentic than reality. Its power depends on its ability to surprise us and trigger the imagination, allowing us to experience the "spark" between two unconnected yet known things. Once a metaphor is shared and put into common usage, it loses its power to enlighten and often becomes a cliché (e.g., "love is a rose"); yet on an unconscious level it can remain deeply influential. Politicians are the masters of such blandishments; consider "morning again in America," made famous in a 1984 TV campaign advertisement for President Ronald Reagan. When metaphors fade from overuse, they don't go away; rather, they remain hidden in plain sight, their etymologies making them discoverable.

Consider the English words "scruple" and "scrupulous." An etymological retracing explains their origins as "moral misgiving, pang of conscience," from the late fourteenth-century Old French *scrupule*, which comes from the Latin *scrupulus*, meaning "uneasiness, anxiety, pricking of conscience," but literally "small sharp stone"—a diminutive of *scrupus*, meaning "sharp stone or pebble." The great Roman orator and statesman Cicero (106–43 BCE) used the term figuratively to mean something causing uneasiness or anxiety, "probably from the notion of having a pebble in one's shoe."[9]

The pebble to which Cicero referred was also the name given to an ancient Roman weight—a small one equivalent to only 1/24th of an ounce. So small a weight becomes a sharp stone in someone's foot, a metaphor for the uneasiness of conscience. What remains of the word's meaning is its abstraction; what is lost is the concrete "lever" that initially gave form to this *feeling of the mind*.

The British archaeologist and linguist Archibald Henry Sayce (1845–1933) wrote, "Three quarters of our language may be said to consist of worn-out metaphors."[10] At first glance, that statement seems over the top—but more than half of the opening of *this* very sentence consists of metaphors, and tired ones at that. It is what American essayist Ralph Waldo Emerson (1803–1882) must have had in mind when he wrote on language in "The Poet":

> The poets made all the words, and therefore language is the archives of history, and, if we must say it, a sort of tomb of the muses. For though the origin of most of our words is forgotten, each word was at first a stroke of genius, and obtained currency because for the moment it symbolized the world to the first speaker and to the hearer. The etymologist finds the deadest word to have been once a brilliant picture. Language is fossil poetry. As the limestone of the continent consists of infinite masses of the shells of animalcules, so language is made up of images or tropes, which now, in their secondary use, have long ceased to remind us of their poetic origin.[11]

What makes metaphors lose their fire and the power to illuminate, in the way that a joke told too often may lose its humor? The jump or spark formed when two unrelated things are suddenly fused alters us—we learn, we remember, and we map that as individuals, triggering further thoughts and flickers of the imagination. Culture goes through the same process as in Emerson's observation on the fossilization of language, a decreased response to stimulus analogous to our five senses' dependence on phenomenal change in order to function. With the power of metaphors to "spark" the imagination, this diminishment, this habituation that affects the senses as it affects language and the imagination, leads me to wonder: What exactly *is* the imagination?

LOCATING THE IMAGINATION

I quoted earlier from Shelley's essay "A Defense of Poetry." Shelley believed the imagination is accompanied by reason, but as a paired opposite. He posed reason and imagination as two types of mental activity, reason being an activity of breaking down and enumerating what is already known and thought, while imagination identifies with those qualities common to existence and nature and as such recognizes the similitudes of things. Put more succinctly, reason analyzes and imagination synthesizes.

This distilled picture of the mind's actions is clear enough and corresponds to our own experience: when we bring two seemingly disconnected things together, *aha!*—a sense of discovery that is credited to the imaginative powers of the mind. When working through a creative problem that relies on the imagination, one works with some medium, such as language, pencil, stone, or whatever, yet this is often accompanied by the peculiar conviction that the imagination functions autonomously, that it acts exclusively in the mind, and that media and method are limited in function to a rather passive, propagative role. We forget, as the American writer and philosopher William Gass (1924–2017) noted: "Things give rise to thoughts, thoughts do not give rise to things, except secondarily as plans for action."[12]

Even in the education of artists and designers, materials/media are often treated as secondary. The heavy emphasis placed on technique, or procedure, by its very nature creates a methodology as opposed to a conversational relationship between author and work. Learning to *control* a medium can limit one's awareness of the imagination's activation through and in materials, media, and even technique. Control *lessens* listening; consider those engaged in the arts who are well versed in methods and instruments, yet never achieve the kind of mastery characterized by a turn to an almost hyperawareness and deep conversation *with* materials and media.

This conversation does not happen within language but within physical media: I press the brush just so and the paint oozes a little more than I thought, and immediately I correct it or—realizing something new has happened—let it go, because now it's better than I was "imagining" it would be. The imagination in this instance is activated *outside* the mind. Robin Evans, an architectural historian commenting on architectural projection systems, saw these supposedly conventional means of transmitting information as rife with unselfconscious distortions and peculiarities and remarked that the imagination seems to lie outside of ourselves—it works only when we activate it by doing something with and through a medium.[13]

Definitions abound for the imagination, but can we establish any facts about its nature? When we try to imagine, we do so introspectively, which makes it more difficult to consider the imagination as dependent on external stimuli. It remains a continual challenge to escape the "I imagine" authority of the "I" mind, so deeply is it embedded in language. We think of explicit reality as "leaving little to the imagination," implying the imagination is closeted in the mind away from reality.

To understand better this aspect of the imagination's nature, I've asked students over the years to test their imaginative powers with a familiar object by closing their eyes and imagining an apple as completely as they can. Then I ask them to draw the intensely imagined apple. The results are remarkable for their consistency and cartoon-like flatness, each drawing a perfectly "unimaginative" apple, almost indistinguishable from all the others. It turns out that the vaunted powers of the imagination disappoint when extracted from the mind as some kind of willful mental operation. Only by either continuing the drawing or looking at and drawing a real apple does the imagination seem to stir or attention to phenomena reemerge.

Neuroscientist and psychologist Michael S. A. Graziano's work on consciousness—which I define as awareness by the mind of itself and the world—offers some evidence-based concepts for why those imagined apples may have ended up being so unimaginative (the emphases are mine).

> When we introspect and seem to find that ghostly thing—awareness, consciousness, the way green looks or pain feels—our cognitive machinery is accessing internal models and those models are providing information that is wrong. The machinery is computing an elaborate story about a magical-seeming property. *And there is no way for the brain to determine through introspection that the story is wrong, because introspection always accesses the same incorrect information.*[14]

Following Graziano's proposition, consider the imagined apple as information; he is saying that there is no way through introspection to elaborate on an internal model of an apple, since the model is fixed—that is, *until something disrupts it.* That "something" is what our senses and body bring to the mind or what we are attentive to—attentiveness being what differentiates consciousness from the unconscious, limbic, and autonomous systems.

Within this frame of reference, the imaginative spark might be the register of when a mental model is disrupted and forced to confront its disjunction with reality. We may think that seeing things such as bruised apples or worm-eaten apples would continually disrupt the internal "apple" model, but that assumes a high degree of acuity in the internal model. If the students' cartoon-like drawings are any indication, the mind may be accessing highly simplistic models of an apple, along the lines of the basic-level categories discussed earlier in the chapter. Of course, if a strange edible living apple arrived in the form of a perfect cylinder, it would cause a jolt. When enough apples become cylinders, the internal model will be modified and the "surprise" will fade accordingly.

In this case, the imagining of a cylindrical apple is *not* an act of finding similitude between two unrelated things, to cite Shelley's description of the imagination. Conflation of cylinders and apples is not similitude. As an *idea*, the cylindrical apple is rather uninteresting; only were such an apple to be grown and harvested would it attain the power to draw our attention. But imaginative fusion hits in full force in a sentence from the 1854 novel

Hard Times by Charles Dickens: "He was touched in the cavity where his heart should have been—in that nest of addled eggs, where the birds of heaven would have lived if they had not been whistled away—by the fervor of this reproach."[15]

To describe a cold, dead heart as a nest of addled eggs is such an astonishing leap, and yet "right" to us. Dickens isn't challenging a mental model; he is igniting our imagination by bringing together two mental models.

The noted scientist and futurist Ray Kurzweil has argued the brain is nothing more than as enormously complex pattern recognition machine or computational organ connected to the world by a network of sensory-motor circuits.[16] This sheds no light on the imagination or, for that matter, on why metaphors stimulate our mind. The architecture of the brain and its structures is understood as an accumulation of evolutionary brains: the reptilian brain is one component of many within the human brain; the neo-cortex (white and gray matter) is another component, the most recent to develop in evolution and the last to develop embryonically as well. Randy Buckner and Fenna Krienen, researchers in neuroscience and psychology, have proposed an intriguing "tethering" hypothesis about brain evolution that may point toward a physiological explanation for the imagination and creativity in humans.[17] They suggest that the massive expansion of the brain "untethered" the tight, hierarchically clear (canonical) circuitry between sensory input and motor output found in most animals—such as when a dog reacts instantly to a squirrel running across the street. More specifically, as the cortex—with sensory input and motor controls mapped into clustered areas—grew, it began to increase rapidly the size of its in-between areas, that is, those areas not wired directly to the sensory-motor circuits. Then something happened that changed the directness of the animal brain. The archipelago of in-between areas—I call them the fuzzy zones—began to network, "talking" with each other more than to the sensory-motor circuits, which directly link to the senses and the physical world.

This "associative cortex" develops late embryonically, which also implies it evolved late, and is one of the key brain features that clearly distinguishes us from our primate cousins. The network is unusual in that the zones demonstrate a strong proclivity for wiring to even their furthest associative neighbors. While the canonically organized sensory-motor network is still fully engaged, this new network also has a robust autonomy, which may help locate our special abilities to work internally: judgment, remembering, imagining the future. Buckner and Krienen also speculate that these associative cortex networks may harbor top-down control networks, that is, networks that control other networks.

Though admittedly an unscientific stretch, I picture in this description an autonomous self-absorbed imagination directing those autonomous networks, unaware of the feedback from the sensory-motor networks, and arcing back to affect deeply the "top-down" mind. All of this corresponds to the way I've experienced the self-sense of "imagination" in so many people.

The "tethering hypothesis" speculates that the rapidly expanding cortex may have interfered with the tight molecular signaling that neurons grow toward. Once a critical distance was reached and the signal weakened, the neurons may have "wandered" and begun networking with other areas.

> The word "tether" is used to emphasize that the expanding cortical plate is tethered to gradients that initially evolved in a cortex with a far smaller surface area. Much as taffy, being pulled apart, thins until it breaks in the middle, the expanding cortical zones far from the strong constraints of developmental gradients and sensory input may become untethered from the canonical sensory-motor hierarchies.[18]

When we are unsure, we dwell in a half-formed, vague, emotionally ambiguous thought/feeling state registered by the effusive use of "uuhmmm" and "kinda" and "sort of." The tethered hypothesis, the wooliness of those networks created between the associative areas of the cortex, is just the sort of architecture to support an ambiguous state of mind.

Other mental traits also point to a "messier" organization of the mind. Cognitive dissonance, the mental traffic jam created when sense inputs don't match or contradict suppositions, reflects an overlay of multiple control systems operating with some independence. For example, the word "red" next to the word "blue" is much more easily comprehended when the words are correspondingly colored than when, say, the word "red" is colored blue. Dissonance puts pressure on us: it is uncomfortable when beliefs and perceptions don't line up. We are motivated to act, and when we are not engaging the habitual gears of self-rationalization in doing *something*, we bring forward the potential to act creatively.

Synesthesia is a kind of "cross wiring" between adjacent areas of the cortex.[19] Sound may register as numbers or, in rare instances, words may trigger smells. More common in children, synesthesia is associated with development and with creativity—the ability to make relations between things that have none. Synesthesia in general isn't that rare and in some respects is a universal trait. The Bouba-Kiki Effect I discussed in chapter 3, in which respondents associate words with shapes (one jagged and the other rounded), has a 95 percent correlation. Though those experiments demonstrated the synesthetic association between sound and image (sound symbolism), it also bears mentioning that other physiological associations have been noted, such as that saying the words "large" and "small" makes our mouths correspondingly larger and smaller.

The most common types of synesthesia also happen to coincide with physical adjacencies in the brain. An image that comes to mind is that of a hyperwired brain with lots of potential for "short circuits" between proximate areas. The potential for nonlinear mental leaps and ability to make metaphors is built into our brain's architecture, given that the adjacencies in the brain don't follow a "logical" plan.

FORMAL AND MATERIAL REVERIES

The French philosopher Gaston Bachelard (1884–1962) identified two types of imagination, the formal and the material.

> One might distinguish two imaginations: that which gives life to the formal cause, and that which gives life to the material cause—or, more concisely, formal imagination and material imagination. These latter concepts, expressed in abridged form, seem indeed indispensable to a complete philosophic study of poetic creation. A sentimental cause, a cause of the heart, must become formal before it can assume verbal variety, before it can become as changeable as light in its many colorations. But in addition to the images of form so often used by psychologists of the imagination, there are—as I shall show—

images of matter, direct images of matter. Vision names them, but the hand knows them. A dynamic joy touches them, kneads them, and makes them lighter. One dreams these images of matter substantially, intimately, rejecting forms—perishable forms—and vain images, and the becoming of surfaces. They have weight, they are a heart.[20]

To Bachelard, the formal imagination is quick to perceive an "image." It is attracted to the novel, to the beautiful, to the surface of things. It is of the eye. The material imagination is mute and almost inert. It is triggered by touch and by smell. "Vision names them, but the hand knows them."

We don't consciously think this way; the two forms of imagination are intertwined. But as we consider this bifurcation, we can feel our way along each axis: one representational, optical, luminary, shaped, attractive, motivated toward beauty; the other eternal, primal, weighted, mineral, elemental, olfactory. As is our nature, we traverse formal imagination first, before the material imagination can be evoked. The detached, dreamlike, omniscient eye to which we attach so much of our imagined powers of imagination lurks just out of reach; it is up to our deep material memories and lived sensations to detect and precipitate an image from the recesses of language and thought. If we dwell in and hold the image (and this is not necessarily a visual image) long enough, we can slowly expand, interrogate, feel, and *materialize* it. This is not a linear process but one of constant return, turning over, reentry. Bachelard characterized it as "reverie," a process "entirely different from the dream by the very fact that it is always more or less centered upon one object. The dream proceeds on its own way in a linear fashion, forgetting its original path as it hastens along. The reverie works in a star pattern. It returns to its center to shoot out new beams."[21]

Imagining in this sense is a building up toward a liminal, epiphanic experience of the whole—the opposite of a reductive, that is, "scientific," method. A new image is fused into being. Invisible substance is given form, as Russian-American poet Joseph Brodsky (1940–1996) does in his essay on Venice (the emphases are mine):

> *The gondola's gliding too was absolutely noiseless.* In fact, there was something distinctly erotic in the *noiseless and traceless passage of its lithe body upon the*

water—much *like sliding your palm down the smooth skin of your beloved.* Erotic, because there were no consequences, because the skin was infinite and almost immobile, because the caress was abstract. With us inside, *the gondola was perhaps slightly heavy, and the water momentarily yielded underneath, only to close the gap the very next second.* Also, powered by a man and a woman, the gondola wasn't even masculine. In fact, *it was eroticism not of genders but of elements, a perfect match of their equally lacquered surfaces.* The sensation was neutral, almost incestuous, as though you were present as a brother caressed his sister, or vice versa.[22]

Brodsky's description of a gondola at night builds its image from material imagination. While eyes want to know and hunger for the sparkle, for the details, for the *image*, they are turned back. We are instead given some brush strokes, actions, weighted bodies, and tactile traces; a material image, itself conveying another—the image of caressing siblings—turns eyes to the carnality of the material image, to its feeling. The picture, the description "gondola," falls aside like a husk, overtaken by an *image*, hovering between the senses, bridging rational consciousness and the unconscious, experienced as haptic emotion. Constructed and calm, it bathes the senses. This, I believe, is what constitutes the epiphany to which Bachelard alluded: an image arises, not as a picture or smell or texture, but as a contextualized and sensual mental event that completes itself in us and alters our outward awareness, a forerunner of perception, in spite of its purely cerebral nature. As Bachelard wrote, "Poetic images of matter do not spring from our instinctual depths, but instead arise in the 'intermediate zone' between the unconscious and the rational consciousness, at the threshold of rational thought, of objective knowledge about the world."[23]

Language is bound to thought, incapable of easy escape without tricks and recruits. Metaphor, analogy, the steady stream of unconscious absorption of experience by the senses, memories, tacit knowledge, the body's embodiment of time and space—all these buoy language, allow it to drift. The intermediate zone, or space of reverie, where corporeal gravity and the mind's weightlessness touch, is a special zone of consciousness in which material archetypes of fire, earth, stone, and the vegetal can slowly press in and well up from below, waiting to be dreamt into being. We are both actors and audience in this play: our witnessing is action enough; anything more—attention, focus, or any other action—and our state of suspended thought, suspended emotion, suspended imagining collapses into worn ruts of habit. The space of reverie is a quiescent room closed off from thought or longing or nostalgia. We dwell there in a state of absence, a receiver off the hook, indifferent and passive. The invisible, unnoticed, passes through our eyes and touches the (retinal?) whorls of our fingertips; ballast utters a silent word and floats upward; black water and lacquered wood disappear into each other's reflections; a sleepy floating cloud of nothing, dust, solidifies an angular geometry—sunlight through a pinhole.

This space of emergence and mergence is fertile ground, tilled and tilled again until earth and air pass through each other. This space is a way station through which things pass into, out of, and through. Sensations, words, and images speak a common language here, no longer tethered to memory, habits of thought, or emotion. They are the raw material of representation, at the service of naming, thinking, imagining, remembering.

Representing things—either with words or images—slows the flow of the raw and undifferentiated world enough to bring things from *out there* to *in here*—to *mind*. And it is the mind that does this naming and marking and differentiating, not our eyes or our noses. Once something is named, it is represented and therefore thought. Images resist naming, but like named things will tend to be remembered and retrieved, stripped of their particularity; we make a habit of living inside the general and the familiar. We barely notice the drive home, numbly filled in with those images we already know—the visual equivalent of speaking in clichés and dead metaphors, our mouths rounding around those phrases knowing full well that they are gaseous nonsense, that "pushing the envelope," for instance, isn't pushing anything. Yet we cannot help ourselves.

We imagine the imagination is free of all this, this hardwired, low-energy mental model that lets us sleepwalk through the checkout line at the supermarket. But is it? The imagination—that is, the ability to make new and meaningful relations and images, original and never before experienced but still influenced by our memories and perceptions—is not immune to habits of the mind. Reverie loosens hardened preconceptions but is not yet imagination.

Imagination is construction, deliberate work; employing chance operations or games—something Mozart toyed with and that became an actual strategy in music composition in the twentieth century, perhaps most notably in the work of American composer John Cage (1912–1992)—serves to free us but is useful only as something akin to a warm-up for an athlete. Too often, chance is misunderstood as "imagination" or "creativity" and fails as both. At best, it yields the surreal. At its worst, chance is an inauthentic, unearned creativity.

Disinterest and inattention are often lumped together, but disinterest—unlike inattention—is a form of awareness. Reverie depends on disinterest, a witnessing without judgment; inattention is associated with dullness of the senses and a mind that's "elsewhere." A state of disinterested attention can propel an observation and bring it to a stage where we can play with it, roll it around without trying to make something do or be something. Possibilities appear when we relax and allow time to slow and expand so attention can fully develop, to entangle thinking with the unconscious. Images take time; they are difficult to make, not simply a matter of an upward roll of the eyes and an "I imagine . . ." The imagination doesn't produce pictures, but rather forms new possibilities from memory, thought, the senses, and material and formal impulses.

The poetic image to which Bachelard refers is not confined to a single modality. An "image" may be formed with words, as a sensation, as something akin to a memory (albeit as a new construction), or even sublimated by the body, as athletes well know. It is said that it takes 5,000 hours to master something.[24] Imagination is part of mastery; we need to imagine doing something, and imagine doing it in so complete a way that when we do it we are simply *following* what we have imagined, and imagined deeply enough to *know* tacitly. Tacit knowledge is material, physical; it is granular, distributed knowing that cannot be explicitly conveyed. Though Antonio Stradivari's violins were made by members of his workshop, each executing one step of the instrument's making, no one but the master understood the exquisite play and adjustments between each slightly imperfect step—tacit knowledge following the deeply imagined Stradivarius into existence. Stradivari never passed on his tacit knowledge; he touched the last of his violins.

The imagination "speaks" in images and "pictures" words—the medium of imagining is ever so elusive, notwithstanding the confidence with which it is felt. Metaphor further complicates things. As the American poet Wallace Stevens (1879–1955) described it, the metaphor is "the creation of resemblance by the imagination. Metaphor creates a new reality from which the original appears to be unreal."[25] Our ability to flip reality on its head is particular to language: we can construct an image with words that displaces our perception of the real thing. The constructed word image is neither a mimetic representation nor description but a felt thought. This is distinct from perception, or from a description of something perceived.

The erroneous belief that words and internal visualization are interchangeable ways of evoking the imagination leads to the insistence that one can conjure an "image" in the mind that needs only to be "drawn out"—the equivalent of, say, writing about an "image." Writing constructs images in its own nonoptical way, a mix of associations, details, structure, metaphor, memories, and feelings. Words easily transfer between the mind and the page. There is no doubt that what I am writing is what I am thinking, even if the words are as formative of thoughts as thoughts are of words. Asymmetrically, the visual imagination doesn't have a ready medium that lives both in the mind and in the world. A stick of charcoal cannot make a drawing in my mind, nor can I imagine in charcoal. Dreamers may protest, but the dream image is immaterial, a pure and unbounded envelopment of us.

Ezra Pound wrote, "When Shakespeare talks of the 'Dawn in russet mantle clad' he presents something which the painter does not present. There is in this line of his nothing that one can call description; he presents."[26] In other words, the image doesn't substitute for pictorial representation but instead presents an autonomous subject into which our subconscious flows and dwells.

That is precisely what Edgar Allan Poe, the American writer (1809–1849), did in his only complete novel:

> Upon collecting a basinful, and allowing it to settle thoroughly, we perceived that the whole mass of liquid was made up of a number of distinct veins,

each of a distinct hue; that these veins did not commingle; and that their cohesion was perfect in regard to their own particles among themselves, and imperfect in regard to neighboring veins. Upon passing the blade of a knife athwart the veins, the water closed over it immediately, as with us, and also, in withdrawing it, all traces of the passage of the knife were instantly obliterated. If, however, the blade was passed down accurately between the two veins, a perfect separation was effected, which the power of cohesion did not immediately rectify.[27]

Such a primordial image has a centripetal effect on our consciousness, sucking associations into itself: blood, surgery, dissection, all the memories and corporeal vulnerabilities that lurk below the surface. The image embeds itself in our minds. It precedes perception; it has bypassed the senses (existing only in words), directly touching our sense memories, our knowledge of matter's nature.

Danilo Kiš, the Serbian novelist and poet (1935–1989), gives us another example. His image of a room is a work of fiction, not a representation.

"Oh," Anna says, "our room"! We glanced around our forgotten room, rediscovering its furnishings, which seemed to have grown darker in our absence. Two beds, old fashioned, made of wood. Two dressers, in which worms had drilled tiny holes and from which a fine pink dust poured, soft and fragrant as a face powder. Marble-topped night tables beside the beds, like the tombstones of children of good family. In the corner, to the right of the door, a couch upholstered in threadbare homespun the color of rotten cherries, a fine old-fashioned couch resembling an upright piano. In the evening, or whenever it was quiet, the springs would burst out into song. Above the couch a lithograph in color: the Mona Lisa. My mother had cut it out of some magazine when the painting was involved in an outrageous theft from the Louvre or was being triumphantly returned, I can't recall which.[28]

As a description, the passage bears little fidelity to the reality of a room. Doing so would need to encompass time and sense widely. What is the light like? What do we see out the window? Is it cold? What textures and stains and old scars cover the walls? Is there dust? What smells, sounds, doors are there? How high are the ceilings? I could go on. But his words—all in a single sentence—can easily shuttle us from details to remembrances, to his mother's interests, frugality, and weakness for some sensational popular scandal, which has long faded.

Movies based on literature that attempt literal recreations of scenes such as the one Kiš describes consistently fall short. They misconstrue words as representations of reality. Any attempt to recreate this room would miss the point—the construction of the fiction depends on a few details (more recollections than perceptions) that allow the reader to witness the author's structured language and the triggering metaphoric associations. We don't know what kind of wood the dressers are made of, nor do we need to, but we know from the wormhole dust that there is a neglectful and tired atmosphere suffusing the boy's life. Were we to fill in all the bits, or even attempt to, we would lose the emptiness between details that allows us, the readers, to bear witness, *to participate in the forming of the image.* Like the gap between the two things a metaphor compares, the image of the room is made more vivid to the interior mind when we experience it in part; the image then becomes receptive to and fuses physical, temporal, and emotional associations. In other words, it becomes narrative.

Narrative (its Latin source *narratio* is ultimately rooted in *gnarus*, "knowing, wise"), is one of our most important tools for making sense of things, for putting spatiotemporal events into meaningful order. Young children form language around a strong desire to fill in gaps between canonical events and deviations from canonical events, creating a story— even a fictional one—to make sense of it all. And we continue doing this throughout our lives. Even if the story is silent and has retreated to the recesses of our mind, it still is our narrative, one that guides our actions when faced with perplexing events, whether they be personal travails or creative challenges.

The room image of Danilo Kiš leaves enough crumbs on the trail to build a sense of time, to recollect and foreshadow events, but also enough distance between the crumbs to call on our own narrative instinct. The *image*, created in the reader's mind, merges the writer's narrative with that of the reader, and the reader's imagination flourishes.

VISUAL IMAGES

While Shelley exalted language, he was less enthusiastic about the arts that dealt with visual and physical matter.

> For language is arbitrarily produced by the imagination, and has relation to thoughts alone; but all other materials, instruments, and conditions of art have relations among each other, which limit and interpose between conception and expression. The former is as a mirror which reflects, the latter as a cloud which enfeebles, the light of which both are mediums of communication.[29]

Nonetheless, Shelley relies on a physical object—a mirror—to construct his metaphorical definition of poetry. It is his unspoken acknowledgment that despite language's ability to articulate the most abstract content— that is, the content "in the mind"—it needs to return through the senses to have meaning and impact and cannot function without reference to embodied reality.

There is an inherent tension between words and images, as reflected in three quotes from three very different sources. In his philosophical novel *The Picture of Dorian Gray*, Oscar Wilde (1854–1900) wrote:

> People say sometimes that Beauty is only superficial. That may be so. But at least it is not so superficial as Thought is. To me, Beauty is the wonder of wonders. It is only shallow people who do not judge by appearances. The true mystery of the world is the visible, not the invisible.[30]

Wilde valorizes the senses, including the visual and images, over thought and the preconceptions that accompany it.

The eminent Italian astronomer Galileo Galilei (1564–1642), in one of his letters on sunspots, posed a straightforward question about the tension:

> So long as men were in fact obliged to call the sun "most pure and most lucid," no shadows or impurities whatever had been perceived in it; but now that it shows itself to us as partly impure and spotty, why should we not call it "spotted and not pure"? For names and attributes must be accommodated to the essence of things, and not the essence to the names, since things come first and names afterwards.[31]

Galileo astutely observed how words could dominate perception, complaining that the true scientific image of the sun was being distorted by the conventional description of it.

And as the American painter Edward Hopper (1882–1957) noted, "If you could say it in words there would be no reason to paint."[32] Hopper reminds us that words are never equivalent to images, that images have qualities that cannot be translated to words.

Yet words have the power to evoke images in our mind; most examples I've discussed are word images, which is only natural given that this is a written work. Observations, too, have the power to trigger our imagination, but in less obvious ways. The power of word images lies in the play between words and perception—words invade the mind and embed imagery there. When we see something, it arrives through the senses and is part of "reality," which is further from thought than language.

Restating what I quoted from Shelley earlier, "For language is arbitrarily produced by the imagination, and has relation to thoughts alone; but all other materials, instruments, and conditions of art have relations among each other, which limit and interpose between conception and expression."[33] He recognized the autonomy of the visual arts—that they establish other relations, unlike the direct ties of language to the imagination. Although his metaphors—light and its clarity in a mirror, or its feebleness in a cloud—are effective in giving us a visualization of his argument, they also serve as a useful demonstration that metaphors are not *actual* images. Were we to substitute words in Shelley's metaphors, we would never achieve the precise thought he sought to convey. At best, we might begin to speculate on the surrealism of the three elements: even if we could conclude that the light is reflected in the mirror and dispersed in the cloud, that conclusion would not be particularly meaningful.

Let's look at a visual metaphor such as the one here, and let's accompany it with the familiar phrase "beauty is in the eye of the beholder" (figure 4.1).

Text and image operate as independent metaphors; they don't rely on one another to "work." When placed together, though, the image subordinates itself to the text, becoming more a confirmation or illustration.

4.1
"My Wife and My Mother-in-Law," a famous optical illusion from the nineteenth century.

Verbal foreshadowing, a term used to describe a tendency of sensory processes to be effected by verbal processes, might contribute to the subordination of image to text. An example of verbal foreshadowing is wine tasters losing their olfactory acuity after describing what they are drinking. More concerning is that when witnesses to crimes are interviewed, their verbal recollections effect and distort their original perceptions. What is said displaces what has been seen or tasted, as if the *image* formed with words competes with and dominates the *image* from the senses.

In his painting *La trahison des images* (*The Treachery of Images*; figure 4.2), the Belgian surrealist René Magritte (1898–1967) presents an image beneath which he includes a seemingly contradictory sentence, raising cognitive conflict. The French phrase "Ceci n'est pas une pipe" means "this is not a pipe."

It is a pipe. It's not a pipe. What is it?

The obvious fact of the painting—that it *is* an *image* of a pipe, and *not* an *actual* pipe—is not immediately "read," since the painting is embedded in a specific artistic or cultural context. Though dissonance is expected from a surrealist such as Magritte, the painting is slippery in that it presents itself as a surreal object—the complicit viewer bends toward that interpretation—but the painting has no inherent dissonance. Its message is perfectly logical, factual. The real dissonance it generates is between our cultural and perceptual preconceptions.

We are rarely aware that our vision is not an objective sense, but rather is guided, modified, distorted by the interaction of thought and visual perception. We see what we *think* we see. "The treachery of images" upsets

4.2
The Treachery of Images, René Magritte, 1928–1929.

the fictional fixity of our vision. Like a good metaphor, it mixes verbal foreshadowing, habits of perception, and the culturally constructed artifice of seeing all together in an unstable brew that manages to provoke plenty of jumps between unrelated parts of the mind.

Another example of a visual metaphor is *instantly* understood: the earth as ice cream cone (figure 4.3). No navigation is required; there is no

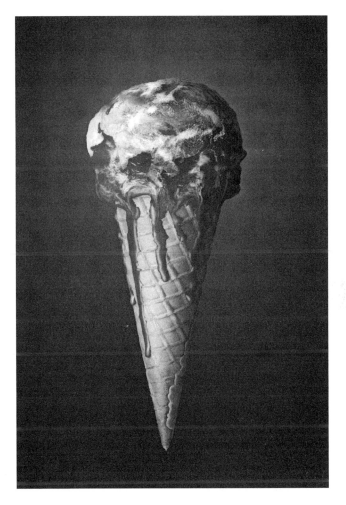

4.3
"The Earth Melting," from an ad campaign by the World Wildlife Foundation.

verbal foreshadowing with which to contend. The carefully rendered dripping cone is an instant cliché, almost immediately emptied of impact—we just get it. The image is so insistent that other associations are subsumed. Metaphor's power, which stems from representing an abstraction of the mind with something concrete of the senses, is dimmed, since the visual "metaphor" conjoins two concrete images—with only a momentary whisper of dissonance.

Nonverbal imagination can take the form of tacit knowledge or tacit image and, unlike a language-based image, is not bound to consciousness. If the conscious mind, thought, lives in the medium of language, it follows that where words cannot go, we cannot *think* to go. The terra incognito, where thoughts cannot penetrate, is where nonverbal images dwell—in the unthought, accessible to the mind through reverie and the oneiric, our dreams. (Note that while I am referring to a nonverbal image, I would differentiate between a deep and meaningful image and what might be called an "I imagine" image—such as the preconceived cartoon apples I mentioned earlier, which are tightly tethered to thought in the form of the word "apple.")

Our memories and dreams are the ore the nonverbal imagination mines and forges into an image. The depth to which we can dig and probe and conjure depends on time; this is slow work, and can only proceed as allowed, not by willfulness.

Bachelard refers to material and spatial archetypes that serve as our depth gauges. When we are in the archetypal attic, the image can be constructed, not written—the archetypal attic gathers our memory to itself and evokes all manner of primal forces and feelings. The association of an attic is with memory, literally. Time and forgetting are bound to the space, the smell, the dim light. Time is stilled: even a short stay brings feelings of mummification. If we keep dreaming of the attic, it will lose its form, its shape, and acquire materiality through the senses and feeling. The raised grain of the raw floorboards catches little wisps of dust and the slightly shrunken twisting knots silently prepare to groan underfoot for the first time in decades.

As the image loses the form and shape of things, it gathers primordial force, becoming material imagination, so present we can touch it. With our attention, the image of the attic loses its formal qualities and gains materiality. This exchange between a "picture" and an apperception of something deeper, outside the bounds of thought, shifts the agency of the imagination from thought to feeling, the former dominated by language and the latter by the senses. In a world dominated by language, it stands to reason that it is material imagination that is overlooked.

Material imagination, in dissolving the formal characteristics of things, is not a reductive act but a construction, a building up, a making. As it is built, the material image draws on material archetypes—the collective memory of matter—and, in doing so, it reanimates and reifies the continuum of culture and self. A material detail, seen in a particular light, elicits this archetypal bond between self and matter (figure 4.4). Whether consciously or not, such material archetypes are deeply felt and draw us into

4.4
A typical brick wall.

their matter. The image of the brick wall in afternoon/morning light isn't particular, nor of a form; it is more of matter than of design. If we focus further, we find "brick" transformed to fired clay, which one might argue is more material than brick. But the material imagination is built from within; we dream the mysterious depth of matter, which is our own hidden depth.[34] Only when we move in our dream state from afar to close up—touch and feel the inner warmth, the hearthness of the clay, and out again to the bricks, the upright stacking, the bricklayer's steady hand—does the material imagination awaken to the alchemy of clay, water, fire, mortar, and hand. Meanwhile, passersby take no notice, but that night in their dreams they see a liquid world turning solid and the ground take flight, waking the next morning in puzzlement.

As strange and inexplicable as this may seem, the dream is simply amplifying (as dreams tend to do) the material reality at hand. All natural matter is liquid in origin. Clay begins as sedimentation in a body of water. When dug, sediments still echo their liquid world. Firing liberates the clay from its repose and the mason's skillful hand brings the brick vertically upward against gravity. And so it is with all materials: the volcanic liquid of granite, the water of a tree's transpiration, the crucible of molten metals or glass, all these are liquids, transformed but still rich with traces of their original state.

Material imagination, in this sense, is material memory.

5 THE IDEA AND THE ACTUAL

The theory of the enactive mind is that of a coupled and dynamic relation between mind and environment in which knowledge is constructed. If we consider the images discussed in the previous chapter as a form of knowledge, we can begin to see how and why the imagination and images are constructed through an enacting relationship between mind and material.

In this chapter, the discussion turns to concepts of the world and how our direct experience of things and materials is both molded by and molds the way we think and create conceptual models of reality.

How we use material changes how we think of it. A pile of dirt is a pile of dirt—but place that dirt in a patch, seed it, and water it, and dirt becomes life-giving, fecund, dark earth. A small branch, with use, might become a dousing rod or kindling for a fire. A taut wire can hang a picture or make a sound. A piece of marble can serve as a floor tile or a cheeseboard. A chunk of charcoal can anticipate a meal or a drawing.

How we think of material affects the ways we use it.

EIDOS AND ENTELECHY

The debate over the contrasting worldviews of Plato and Aristotle still resonates and influences our view of materiality more than two millennia after they lived. Plato, in some eleven of his dialogues (including most significantly *Meno*, *Republic*, *Symposium*, and *Timaeus*), wrote of ultimate truths that lay beyond matter and that manifested in image/idea.[1] He called such an image or form an εἶδος (*eidos*, the Greek word for "visible form"), and his so-called "theory of Forms" that appears in so many of his dialogues posits that nonphysical "essences" (forms) represent reality in its most accurate, complete state. Reality, according to Plato, can only be grasped by sweeping aside all that veils and obscures, namely materiality, flux, change, any state of "becoming" or transience.

Aristotle, in *Physics*, *Metaphysics*, *Nicomachean Ethics*, and *De anima*, wrote of reality as defined by material's potential to become actual. He coined his own term for this, ἐντελέχεια (typically *entelechy* in English), best defined as the actualization of form-giving forces.[2]

Consider the definition of a circle. Platonic reasoning would dismiss all attempts to draw a circle with compass and pencil as imperfect and not real. Plato might call such attempts material approximations that obscure true reality—namely, the figure/image of all points on a plane equidistant from a point on that plane. By contrast, Aristotelian reasoning would involve looking at physical reality's potential to form something actual, such as a stone rolling down a hill or a drop of water sending out circular ripples, as a way to understand a circle as arising from matter.

The relationship between *eidos* and *entelechy* is rich and contradictory. Consider two vessels: a jar from an Egyptian burial chamber (figure 5.1) and a ubiquitous New York City water tower (figure 5.2). The Egyptian vessel contains no volume, since burial custom required only the form/image to accompany the deceased to the afterlife. The material is inconsequential—an example of *eidos*. In the New York City vessel, though, material matters: steel bands resist the water's weight that presses against the wood slats, and the gaps between the bands decrease as the pressure from the weight of the water increases toward the bottom, actualizing the form-giving forces—in other words, *entelechy*.

Another example can be seen in two lives remembered. A small sarcophagus containing a snake skeleton (figure 5.3) bears a representation of the snake on its lid. Thus, image displaces matter—*eidos*. By contrast, old hammers (figure 5.4) record a lifetime of work by craftsmen, their wood handles worn into strange, twisted form by countless hours of use. Life itself was the form-giving force that actualized itself in these handles—*entelechy*.

5.1
Jar (top row, left) known as "Opening of the Mouth," from an Egyptian burial chamber, c. 2200 BCE. Photograph © 2019 Museum of Fine Arts, Boston.

The dichotomy between *eidos* and *entelechy* can be observed in every field, from science to politics to design. In architecture, it is as ancient as it is current. The meditations on the corners of buildings by the German-American architect Ludwig Mies van der Rohe (1886–1969) are a case study of *eidos* and *entelechy* in tension. At the campus of the Illinois Institute of Technology (IIT), Mies resolved to work with the great modern American construction element—the I-beam. He strove to create architecturally pure corner conditions, but using I-section columns at the corners presented a complication: their unidirectional nature prevented a simple symmetrical resolution, the stripping down to what he called "almost nothing" that Mies sought in his architectural work.

The first of his IIT buildings, the Library and Administration Building of 1944, was resolved with a frank assemblage of I, L, and T sections of steel, a kind of *entelechy* that expressed the logic of steel construction (figure 5.5). In subsequent Mies buildings, the steel I-beams were encased in concrete; the smaller mullions march over the entombed I column, presenting

5.2
Water tower, Broadway and Spring Street, New York City, Bernd and Hilla Becher, 1979.

5.3
Coffin for a serpent mummy, Egypt, c. 745 BCE. Photograph © 2019 Museum of Fine Arts, Boston.

a pure corner condition—or at least the *image* of a pure and symmetrical corner condition (figure 5.6). This resolution, in which representation masks actuality, is *eidos*-like or eidetic in nature, the little mullions—like the serpent on the cover of the sarcophagus—presenting the image of the encased I column.

5.4
English craftsmen's hammers from Sheffield, late nineteenth to early twentieth century.

EXPERIENCE AND CONCEPTUALIZATION

René Descartes considered the senses to be limited, insufficient to grasp the nature of the world. In a famous thought experiment, he argued that while our senses can inform us of all the various characteristics of, say, a piece of wax—smell, texture, appearance, and so on—those same senses no longer figure in the same way when that wax is held to a flame and melted. We

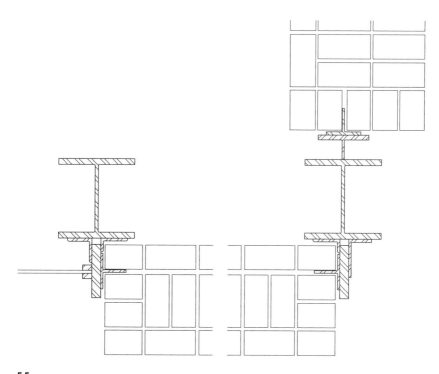

5.5

Steel corner section from plan for Library and Administration Building, Illinois Institute of Technology, Mies van der Rohe, 1944.

know it is still the same piece of wax not *because* of our senses, but *despite* them. Our mind puts our senses aside as it conceives of the wax. As Descartes wrote, "And so something that I thought I was seeing with my eyes is in fact grasped solely by the faculty of judgment which is in my mind."[3]

What has come to be known as the *wax argument* is a simple demonstration of the fluid continuity of mind and sense. One defines the other: what is thought is what is experienced, and what is experienced is what is thought. Similarly, the *Weltanschauung*, or worldview, that is *eidos* and *entelechy* informs our reality: we project through and past material to the meaning it holds for us, while the same material is presenting to us the nature of its physical reality. Reality in this scheme of things is malleable. In its more stable form, it is understood within a conceptual framework open

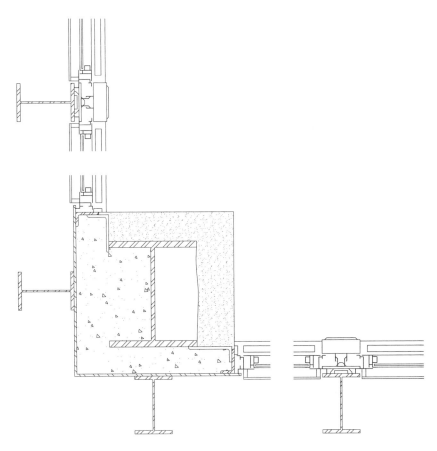

5.6
Corner section from Lakeshore Drive Apartments, Chicago, Mies van der Rohe, 1951.

to limited modification by perceptions in its more intense moments, reality is foregrounded perception—raw experience accompanied by feeling, with preconceptions and conceptual frameworks pressed into the background.

The world as we find it is an infinitely rich complex of sensation. This reality, however, is never "true" or absolute. We live its flux and flow, experiencing time and space as a contracting and dilating continuum, all the while paradoxically conceiving of it as an immutable and absolute construct.

Metaphor sparks the mind when we define abstract notions with concrete terms. The inverse also holds true; science defines concrete things as part of an abstract framework. We select and attend to it within the limits of our senses: our starting point is that if something cannot be seen, it doesn't exist. But then we extend our abilities by inferring, inducing, deducing, imagining, conceiving: I know you're behind the wall because it's your favorite spot for lunch; it's noon; I saw you walking toward it earlier; and I can see the squirrel in the tree chattering at someone. In short, we extend our capacity to sense by other than sensory means.

After a week, I just need to look at the time and I know you're there behind the wall. The squirrel, the wall, and all the rest of it recedes, eclipsed by my efficient conceptual construct, which frees me to think about all sorts of other things. We depend on this instinct to conceptualize the world in which we are enveloped—to inhabit it, to make a habit of it, literally "to take hold of" it. We take the sensations of it (the concrete) and conceptualize it (the abstract).

The history of ideas and discovery and invention is a story of habits of the mind being overturned by force, reason, evidence, and belief. This plays out at all scales—from the single soul who says "I changed my mind" to generational battles fought over ideological belief systems. Habits of thought and preconceptions have their place, of course. Without them, we would be unable to focus our attention: everything would overwhelm our senses; our minds would be taxed dealing with an endless state of confusion. Imagine driving a car without any of the necessary preconceptions (such as conceiving of the road as continuous and made of solid material that won't yield to the weight of vehicles). You might as well drive with your eyes closed.

What our senses tell us can be overridden by strong concepts, and the ones at the heart of a society's conceptual framework carry tremendous inertia. Since before recorded history, for example, the night sky has been a palimpsest for the exchanges between experience and conception; how humans mapped it has always been an uneven mixture of both.

We can be certain that for at least the past 5,000 years humankind has shared the same phenomenon of fixed and moving points of light, and our will to give it order—expressed in the naming and structuring of constellations—developed slowly over time. Out of an unknown prehistory, the Babylonians developed the oldest recorded ordering system of the stars. Successive civilizations continued to adopt and add to earlier orders: the Babylonian system was adapted by the ancient Greeks, and medieval Europe continued to follow it. By then, star maps typically featured writhing bodies of the various constellations, all but obliterating any reference to the appearance of the stars that were ostensibly the subject matter (figure 5.7).

By providing measure in the form of a known (animals and people), these representations were essential to giving meaning to the mysterious, starry sky, and the heavens were transformed into a firm, immutable order that undergirded the religious and astrological governance of everyday life.

5.7
Star map, "Coeli stellati christiani hemisphaerium posterius," from *Harmonica macrocosmica*, Andreas Cellarius, 1708.

COSMOLOGICAL CONCEPTS AND OBSERVATIONS: THE EXAMPLE OF KEPLER

Social cohesion depends on a shared conceptual framework. Creativity, which by its very nature challenges and overturns concepts, has often been met with suspicion and outright suppression. The history of cosmology is a story of creative efforts of discovery clashing with established concepts and orthodoxy. The ancient Greeks were keen observers and developed quite sophisticated models, including heliocentric ones, that were at least influenced by their observations. Later, during the medieval period in Europe, a purely theological basis for cosmology took hold, and astronomical observations more or less ceased.

It is no coincidence that conceptual frameworks, whether secular or based in religion, are far less tolerant of change—especially change based on empirical evidence.

Johannes Kepler (1571–1630), a German astronomer who figures prominently in the scientific revolution of the seventeenth century, entered the scene well after Polish astronomer Nicolaus Copernicus (1473–1543) had published his heliocentric model at the end of his life.[4] Kepler's early astronomical work was a search for the order of the solar system from efficient design, which evidenced the existence of the designer God. His *Mysterium cosmographicum* of 1596,[5] the first published defense of the Copernican system, proposed a geometric basis for the relationship between planetary orbits.

On July 19, 1595, he had an epiphany while teaching. He was demonstrating the relation between two circles by inscribing a triangle in the outer one to determine the inner one, a nesting of circle within triangle within circle. To his astonishment, he realized this happened to describe exactly the relationship between the orbits of Saturn and Jupiter. Using regular polyhedra in a similar fashion, Kepler developed a nested geometric proposition to describe the orbital relationship between all the known planets. Amazingly, it sort of worked, thanks in part to imprecise astronomical data of that time (based mostly on the observations of the ancient astronomer Hipparchus) providing license for some shoehorning. The data did not fit the concept exactly, but it was enticing to have an idea that seemed to offer enough "proof" as well as a "model" for the universe.

Kepler's *Mysterium* was wrong, but it was key to his later discoveries. Kepler's genius was in recognizing relations between unconnected things such as Saturn, Jupiter, and nested circles, but it was also his ability to rearrange concepts, to loosen them. The *Mysterium* was based on circles describing orbits, a concept Kepler worked with for decades—before he eventually discovered elliptical orbits by paying attention to data he had had a hand in producing.

What the *Mysterium* propelled forward was the notion of a mechanical/physical explanation of planetary behavior that cleaned away the encumbrances of earlier Greco-Roman and medieval cosmology, such as the many epicycles (suborbits) that were added to bring models into congruence with observations. In comparison to that sort of Rubik's cube cosmology, Kepler's *Mysterium* was geometrically and conceptually clear and matched observed data about as accurately as earlier models. The precision of the *Mysterium*'s geometry might have influenced Kepler to seek better observations, without which the disagreements between observations and his model would remain unresolved.

Since telescopes had not yet appeared on the scene, observations were made with sighting instruments and the naked eye. Tycho Brahe (1546–1601), the Danish astronomer, had the greatest observatory of his time; with his massive instruments (sextants, quadrants, and armillary spheres) he was able not only to make very accurate measurements but was committed to a regime of constant measurement, assembling the best astronomical data of the period—which, unfortunately, he secreted away. By chance Kepler, who joined Tycho as his assistant, was tasked with making measurements of Mars, which at Tycho's untimely death Kepler "usurped" (as Kepler himself put it).

Mars, it turned out, had been subjected to the most exhaustive and accurate measurements, which made it possible to discern precisely its orbital characteristics. Tycho's instruments were able to achieve a high degree of accuracy because of their size; since it was all about aiming at and reading off angles, the larger the instrument, the more accurately the angles could be read. Some of Tycho's instruments were as large as small buildings, and it must have been quite a physical undertaking for Kepler to make measurements.

One can imagine that, in aiming at the red planet again and again and again over time, Kepler absorbed more than numbers. Tacit knowledge comes from making, from physical action, and this was just that: if high divers slowly acquire a deep sense of where they are as they spin rapidly toward the pool, wouldn't Kepler's very physical observations have slowly given him a tacit *sense* of the motion of Mars relative to Earth? Put in the terms of extended versus enactive mind introduced earlier, it would seem an instrument extends the mind, while a physically embodied experience is more enactive. The giant instruments Kepler worked with were ostensibly an extending of the mind through measure, but could the physicality of working the instruments have constructed a kind of knowledge characteristic of an enactive mind? Could Kepler, in those countless hours of measuring and moving, have begun to experience the spatial depth of the planetary orbits?

Once Kepler had the measurements in hand, hundreds of pages filled with numbers, he spent more than six years trying to make sense of Mars's orbit. He moved from one supposition to another, using an egg-shaped ovoid orbit as a heuristic figure to help fit the data while still expecting that the actual orbit would be something fitting a pure—that is, "divine"—geometry.

Sixty-five years earlier, Copernicus had revived the heliocentric model, but notably made almost no observations to confirm its veracity. The cosmological challenges he took on were to improve on the mechanical/geometrical puzzle of epicycles and circles and their mathematical relationships that had been developed by the Greco-Roman astronomer Ptolemy (c. 100–170). Copernicus was tweaking a conceptual model and not working from empirical evidence. The *eidos* of cosmology had not yet made room for *entelechy*.

It took a trove of precise observational data to nudge humankind toward a new cosmology. Before Tycho, no one had made precise enough measurements to provide observational evidence for elliptical orbits. And since only Mars had a sufficiently eccentric orbit to be recorded within the accuracy of Tycho's instruments, it was highly fortuitous that it became Kepler's assignment. We know that from these data Kepler plotted out the plan of the orbit of Mars and was aware of its slight eccentricity. But he couldn't define it as a geometric construct; it was a set of points.

Creative work, regardless of discipline, is full of such "wrongness" that nevertheless carries the "right" within itself. First appearances, particularly at the beginning of a creative problem, are powerful, there being little to challenge them. In my experience, one thing is consistent: regardless of how one begins, the *whole* of it is contained in that first foray. Creativity is as much about clarification as it is about release from assumptions, a letting go so that possibilities can be allowed to enter.

It took further creative work over quite a number of years before Kepler found his answer and could release that whole. He knew the ratio between the radius of the circle and the short axis of Mars's orbit (nested in the circle) to be 1.00429 (from direct observations) (figure 5.8). By itself, this meant nothing; it was just a data point. But he happened to measure the angle between the position of Mars from the center of its orbit and the Sun (CMS) and found it to be 5.3 degrees. By chance—or owing to the power

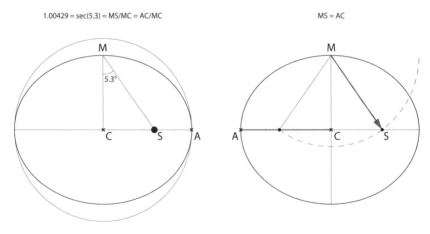

5.8
Mars orbit overlaid on a circle, after Kepler.

of the unconscious—he also happened to check the secant of the angle, or ratio between lines MS and MC. It was 1.00429.

As he wrote later, "I felt as if I had been awakened from a sleep."

Kepler stumbled on the proof that Mars's orbit was indeed a true ellipse. Every ellipse is defined by a simple proportional relationship: AC/MC = MS/MC, which in Mars' case was 1.000429. To define an ellipse physically when the major and minor dimensions are known, one has to determine the foci. Simply draw an arc from the end of the minor axis (M) that is one-half the length of the major axis (AC). The foci are where that arc crosses the major axis. (S). Since AC equaled MS in Kepler's plot, this could only be an ellipse.

Until Kepler, much of Ptolemy's cosmology had managed to survive intact. The Ptolemaic universe was born from an effort to unite a conceptual framework (assumptions) and observational data (experience). Like any creative process, its slow transformation was full of blind alleys, dead ends, and wrong assumptions. Based on Aristotelian principles—namely, that orbits must be circular and motion uniform (which even Copernicus had accepted)—it required the introduction of extra pieces of inventive geometry in order to approximate the observed data: epicycles (a small circle, the center of which moves around the circumference of a larger circle, used to account for the periodic irregularities observed in planetary motions); deferents (a circle centered on the Earth around which the center of the epicycle was thought to move); and equants (an imaginary mathematical point used to explain the observed changes in a planet's speed when it is closest to the Sun). Even though epicycles were completely wrong—they were invented to keep the idea of circular orbits intact—they did plot out noncircular orbits, even elliptical orbits. They were a wrong mechanism based on a wrong assumption that nevertheless pointed toward the truth, carrying bits of the answer: epicycles share characteristics (but not geometries) with ellipses, such as consisting of two centers (or foci), a constantly changing relation between long and short axes, and a kinetic construction that employs both center points.

It took Kepler most of his life before he could finally let go of the circular orbits he loved most—despite all the evidence pointing him away—and open himself to elliptical geometry. It was like swinging through vines: one flawed idea could carry him only so far before he would have to drop it in favor of a new one, grabbed hold of with no assurance of forward progress and almost a guarantee of difficulties, obstacles, and surprises ahead.

Copernicus was bothered by one of Ptolemy's geometric inventions in particular, the equant, which was added into the mix to fit with observations that the planets appeared to be moving at different speeds over their annual orbit. The equant was a theoretical location from which a planet could be observed to move at a constant angular speed—thus preserving the "sacred" idea of constancy of motion. Copernicus and others thought this the weakest part of Ptolemaic cosmology; the equant was eliminated in the Copernican universe, but additional epicycles were added to conform to data. Kepler began with the Copernican model but saw the epicycles as hopelessly unable to match his new and precise observations. So he got rid of the epicycles and, in doing so, found himself plotting orbits that didn't center on the Sun but were offset. He recycled the discarded equant and in doing so laid the foundation for his three laws of planetary motion: (1) the paths of the planets that orbit the sun are elliptical in shape, with the center of the Sun located at one focus (the law of ellipses); (2) in the planet's orbit, an imaginary line drawn from the center of the Sun to the center of the planet will sweep out equal areas in equal intervals of time (the law of equal areas); and (3) the ratio of the periods of any two planets is equal to the ratio of the cubes of their average distances from the Sun (the law of harmonies).

Once the orbital geometry was solved, it was clear that the equant was in fact the other focus of an elliptical orbit around the Sun (the Sun located at the first focus) and that the constant angular motion "seen" from the equant did, in fact, appear to hold true with elliptical orbits. (It was considered a fact until the nineteenth century.) More importantly, it guided Kepler to the second law, which is similar in some ways: instead of equal angular motion being swept over equal increments of time by a planet from the focus/equant, an equal area of the orbit was swept by the planet over equal increments of time. The equant was a wrong idea that supported a wrong assumption; it was invented, discarded, and reused, ended up wrong again, but nonetheless was the critical ingredient from which Kepler's second law was derived. It was a lie that carried the truth.

Astronomy by nature is nonmaterial: its subject matter lies beyond reach, and sight was extended only when Galileo looked through a telescope. Absent any physical facts, it is little wonder that cosmology was driven by philosophical, theological, and geometric theories and ideologies—each influencing and stabilizing the others, making it very difficult to question circular orbits. As we have seen, Kepler broke through, armed with new data and a determination to have the data reveal the design. In his obsessing over the data—both the measuring and then the mathematical work— and perhaps as a result of it, he managed to spatialize the solar system, to *experience* it.

When three centuries later Albert Einstein called an idea of Kepler's "pure genius," he wasn't referring to elliptical orbits or the other laws; it was how Kepler determined the orbit of the earth. To determine Mars's orbit, Kepler first needed to know Earth's orbit relative to the Sun. To know that, he needed another fixed point in addition to the Sun so he could triangulate Earth's location in space. But there weren't any available, all the planets being in motion.

Kepler's "pure genius" was to relocate the point of observation to Mars. Knowing Mars's orbital period, he was able to fix Mars (relative to the Sun) using the data from the same Martian "new year's day" of each Martian year. Thus, Mars became a "fixed" observatory, with the Earth being measured relative to the Sun. It was not the first time Kepler had made such a leap into space. Earlier in his life, in 1608, he wrote a novel titled *Somnium* ("The Dream"), which has been called the first work of science fiction.[6] In it, he imagines a trip to the Moon and describes how Earth would look if viewed from there.

Kepler was not only able to move spatially throughout his model of the solar system, but also sought to understand it physically, to get at the motive forces driving the planets around the Sun. This was in stark contrast to his contemporaries' theories of a dematerialized, abstract cosmos constituted of *quinta essentia*—the aether conceived to fill the universe beyond the Earth that was thought to be a pure fifth element (after earth, air, fire, and water) and, by virtue of being unknowable, didn't need to conform to any known physical behavior. Kepler was the first to draw the orbits of Ptolemy's system so he could take a close look at what the geometry of epicycles actually described—and in doing so discovered an absurd, nonrepeating, pretzel-shaped looping figure.

Apparently, no one had ever bothered to check, or simply no one was concerned—since the *quinta essentia* was free from physical laws.

Kepler may have been an idealist in spirit, but he was a materialist at heart. He was the first astrophysicist, thinking of cosmology as a physical science that worked with physical laws. He wanted to understand the elliptical geometry of orbits as *entelechy*—the actualization of forces—and ascertained most of the nature of gravity: the inverse proportion of its force over distance, and its proportional relation to mass. He used an awkward analogy of a single-oared rower on a circular river, the oar propelling one faster and slower depending on the angular position of the blade, to explain how planets move at various speeds (he later jumped to a more satisfactory analogy using magnets). The analogy of the rower demonstrates Kepler's empathetic reasoning: he enlists an embodied experience to feel the physics behind the numbers, to get at an answer to the question of motive force.

We can't help thinking of the Sun as rising in the East, rather than ourselves as spinning eastward to encounter a stationary Sun. It is ironic to think that it was daily experience that got it wrong and led to a geocentric model of the universe. The persistence of geocentrism, so buttressed by lived experience, was prolonged by the hunger for meaning, which historically manifested itself in religious belief. The binds of religion came naturally, providing a narrative structure to fill the immense celestial void. And once experience acquires a narrative, that experience is thought, represented, and symbolized into a stable structure resistant to change and susceptible to what the English philosopher and scientist Francis Bacon (1561–1626) referred to as the "idols of the tribe"—embellishments, exaggerations, and distortions that, over time, get commingled with facts and perceptions.[7]

That Kepler got so much right, given the tools available to him, is remarkable. Even more remarkable was that he was able to break from being in the thrall of established celestial "truths," including those of astrology, which so diverted attention from the planets. Just as those who read tea leaves look at the leaves only for their symbolic content, ignoring their physical properties, few before Kepler looked at the cosmos for something other than its meaning and symbology (hence astrology). Kepler, though, transformed the metaphysical universe into a physical structure; he was the first to lay out the full implications of thinking of the Earth as a mere sibling among the planets in our solar system.

SCIENCE OF THE CONCRETE

Although they got the order of the heavens wrong, our early ancestors were extraordinary observers, diligently measuring and tracking the movements of the Sun, Moon, and five visible planets, and even predicting eclipses and alignments. This level of interest and intensity of engagement was widespread; marks, monuments, and structures made this information intrinsic to life. While astrology might now be regarded as a pseudo-science that retarded the development of astronomy, it was nevertheless a systematic structuring of experience.

In his essay "The Science of the Concrete,"[8] French anthropologist and ethnologist Claude Lévi-Strauss (1908–2009) compared "primitive" systems of ordering with science, and demonstrated, with numerous examples, the former's scientific attributes. The fundamental impulse of science is toward order, and so, too, were the efforts of earlier societies. They sought to make order based on what was known to the senses, order of any type being better than chaos.

Order came first from identifying, naming, and differentiating *this* from *that*. It is a widely held assumption that primitive and early cultures paid attention only to what was useful (practical) to them, but this is refuted by anthropological studies that reveal the studying and naming of both useful *and* interesting (theoretical) things.[9] The curiosity of so-called primitive peoples about their physical environment is no less conceptual than that of our modern society.

Concepts are associated with a degree of abstraction—at a glance, "tree" may seem more conceptual than "birch" or "oak." But having many more specific words to describe things doesn't limit the concepts those words contain; in fact, identifying and naming many more varieties of things based on experience might well be considered far richer than identifying few varieties under the aegis of a concept. Far from regarding the nomenclature, taxonomy, and mythical thinking of early cultures as a simplistic early phase in the evolution of science, Lévi-Strauss thought of this as a parallel "distinct mode of scientific thought," one roughly adapted to perception and imagination, the other, modern science, being at a remove from it.

Western science begins with a hypothesis and proceeds to build a model to fit. By contrast, "concrete science" accepts the entirety of experience and attempts to structure an encompassing narrative. In this context, magic, witchcraft, and mythological thinking even today are needed structural components that anticipate, but do not yet have access to, knowledge. They pay close attention to nature, attempting to relate disparate things and to determine causes. As Lévi-Strauss explained:

> This preoccupation with exhaustive observation and the systematic cataloguing of relation and connections can sometimes lead to scientifically valid results. The Blackfoot Indians for instance were able to prognosticate the approach of spring by the state of development of the fetus of bison which they took from the uterus of females killed in hunting.[10]

Systemizing insights and relationships that are presented to the senses and experience is a difficult undertaking. For early societies, causality was the glue holding everything together. If this bird's beak can be related to that internal organ, it is a valuable relationship—even if only to make order. The need for order led to a need to be deterministic about everything, including things that had less causal connection than bison fetuses and the time of year. Magical thinking fills in the missing parts of the deterministic universe. Sacred things are sacred because of their place in this structure, and were this to be disturbed the entire universe could come apart.

In this context, naming the constellations was the effort of "concrete science" to extend terrestrial order to the sky. Order was projected onto the celestial sphere in the form of a gigantic simile: animals, travelers, heroes were called upon to give form to its exotic and unknown pieces. Astrology was overly deterministic so as to keep order in the cosmos.

Over time, a people's arrangements and structures will evolve; those that confer a benefit will be remembered and passed on, strengthening the order, and those that are detrimental will tend to fall away. The Neolithic technological revolution exemplifies this: an early epoch of careful observation and ordering, leading from discovery to discovery, with agriculture, cloth, pottery, and cities all achieving a very high degree of advancement in relatively short time. The development of ore smelting is often cited as evidence that these discoveries were not by chance, since no matter how much copper ore is accidentally thrown into a fire, nothing comes of it. The only way copper metallurgy could have been discovered was as a byproduct of the firing of glazed pottery, which demands persistent experimentation, keen observation, and insights into causality. All these successes came from the events and structures of concrete science—a demonstration of its power to invent and make discoveries.

6 PROJECTION

In previous chapters, I've discussed the confluences between the mind and material through various lenses, including anthropology, language, and philosophy. The next several chapters focus on the instruments, tools, procedures, and surrogates involved in the transformation of material and physical reality, especially by those who work closely with a medium, such as artists and designers. First, I look closely at the development of perspective and other forms of systematic projection and examine how painting, sculpture, and architecture have been deeply influenced by such techniques, which in turn have informed how we work with and conceive of materials. Far from being merely a form of neutral information conveyance, projection itself is implicated in the ideas and realities of our surroundings, despite its almost invisible presence in the making of images and things. A painter paints, and how projection revolutionized and then continually influenced representational painting is of lesser interest to most historians, because of the tendency to discount the role of instruments (such as projection) and media/material in the creation of a work. The hand, in particular the artist's hand, more often than not is assumed to be the sole agent guiding the working process.

What could be more authentic than our signatures? Even though our hand is holding an instrument (a pen), our signatures are considered a direct, unmediated act of our selves. A signature made with a finger—even though it is a more direct transmission—could reasonably be viewed with suspicion. In this example, "the hand" is a tool-holding hand. Tools, for those who work with material, are seamless extensions of the hand (and mind). Each discipline traditionally limits the allowable degree of extension: for instance, a woodworking artist is still "authentic" using power tools, but not if the tools are handled by assistants; visual artists and architects have "authorship" even though they often rely on systematic geometric projection, such as perspective, in their work.

The extensions of the hand, in particular projection, have changed how we work with material and have altered our apperception of reality. Projection has given us the power to remove ourselves from direct experience, to create representations of reality and place ourselves within those representations. How such representations are materialized and how material is affected by and in turn informs mental images and cognizance become the focus of this chapter.

THEATER

It may seem unusual to begin a chapter on projection with the subject of theater. However, when the first audience—what had originally been the chorus—gathered together in order to experience a representation of life as portrayed by actors, the concept of projection was born. German philosopher Friedrich Nietzsche (1844–1900) examined this great reimagination of the individual's relation to the "all" in his early work *The Birth of Tragedy*.[1] As Nietzsche describes, two ancient Greek gods who were sons of Zeus, Apollo and Dionysus, embodied two sides of the relation of the self to the world: the former a calm individualism contemplating the world as image; the latter a magical primal immersion into nature. Nietzsche proposed that the Dionysian and Apollonian are twinned urges to Art: Dionysus to music and Apollo to the plastic arts (the term used to refer to those forms of art that involve the physical manipulation of materials that can be shaped or carved; from the ancient Greek word meaning "to mold"). The clash between these urges, in Nietzsche's view, led to a continual rebirth of creative life in ancient Greece and a struggle between order and passion.

Apollo was the god of reason, harmony, and order, and thus the Apollonian is based on logical thinking and rationality. Dionysus—the god of unrestrained irrationality and chaos, of fertility, nature, and wine—was celebrated with dance and music, a kind of "intoxication" as a means to reach a state of ecstasy: literally "the act of standing outside oneself." The Dionysian is that which appeals to instinct and emotion; the worshippers of Dionysus sought to dissolve themselves into the "primal unity." Although little is known about the origins of acting, the dithyrambic chorus—a more formal version of the worshippers' "drunken dance" in the Dionysian festivals—arguably was a key development. Originally there were no actors in the theater, nor was the idea to "act" as we understand it today; there was only the chorus. Professor Mark Damen summarizes how many scholars believe the role of actor emerged from ecstatic worship.

> At the highest pitch of the celebration, it was believed that worshippers left themselves, as the god entered their bodies, and they could then perform miracles and wondrous acts beyond normal human capabilities. That the "impersonation" of a human by the god bears some resemblance to what an actor does in performing a drama has led many a scholar to assert an evolutionary connection between this sort of worship and the performance of drama.[2]

Dionysus was a representation of a fundamental urge to become one with the all, to dissolve all distance between the self and the vitality of nature. This Dionysian urge was not without danger: lust and cruelty and madness could sweep aside all civility and become a threat to Greek order. The Dionysian urge, though, was authentic as a reaction to the artifice of the calm and contemplative Apollonian spirit. Apollo the seer, the god of light and of the plastic arts, was the embodiment of the urge to appearances, which veiled and protected one from reality, a sort of dream state in which the individual could calmly navigate through the tempestuous storms of reality without fear. The statues of the gods played their role, protecting the individual by their appearance. Against the dangers of the "out of control" Dionysian urge, one can recognize the limits of the Apollonian: a trust in the illusion of appearances and in the supremacy of the principal of individuality, a complete internalization of the world.

In Nietzsche's interpretation, Greek tragedy is born of a unification of these two urges—a unity of opposites, which is the nature of Art. The dithyrambic chorus, purely Dionysian in nature, was itself in a state of ecstasy, outside of itself. The hallucinating singing and dancing chorus cannot *represent*; it *is*. Yet the chorus is already a step away from the dissolved mass of worshippers; it is a prototype of what we understand as an image. The spectator remains outside, is reminded of but unable to participate in the enchantment of the chorus—until, that is, an actor is created, an actor being not the thing itself but a representation, an appearance, an Apollonian image of the Dionysian chorus, what the chorus feels and envisions in its transformed state. The audience then can "see" the actor as image, voice, appearance, and can hear and feel through the agency of the chorus. The chorus is in the all—there are no selves, and it sings and moves in unison—while its dreamt image, the actor, speaks in a disembodied third-person narrative. The spectator then experiences the all through the fusion of image and feeling; the chorus births the actor, a representation of the felt pain of reality, as a form of art. The chorus has invaded the Apollonian world; the image of the actor cannot be regarded as an image of contemplation, but as a hallucinated part of the authentic and writhing unity of the chorus, the birth of drama.

The space between spectator and play, chorus and actor, remains the space of individuation. The spectator experiences the play in isolation in the theater. This, historian Alberto Pérez-Gómez has argued, is *the* space of architecture in that it is a space that traffics in both the meaningful and the measurable, enveloping and projecting, the fluctuating urge to collapse and expand—the former into the felt totality and the latter into the dream world of images.[3] This characterization reminds us of the essential mystery of space; it is why we recognize our self in the gaze of another, and why the panoptic eye of surveillance, which strips space of its depth, is so much more an affront than to privacy alone.

As Walt Whitman wrote, "All architecture is what you do to it when you look upon it."[4] The extromissive eye extends the act of "looking upon" to the concept of a visual perception in the form of rays of light emitted by the eye; though scientifically false, this theory of vision aligns with the magical nonoptical transactions between subjects across space. What Shelley describes as the dimness of the material world might in spatial terms be likened to a fog that reveals thickness and dimension, in contrast to the bright transparent and immaterial light of day that hides the conductive capacity of space. An example of such materialized space is nocturnal space, which still manages to hold us in wonder and in heightened awareness of its envelope. At night we can feel crepuscular eyes on our back, and as we move about, the fluctuations from intimate to immense bind image and matter.

Space is neither medium nor language but, like language, is both of mind and in the world. The space of Greek tragic theater is a model of the imagination: the individual spectator's experience of raw life, in the dithyrambic chorus, and the images it produces (the actor). This imagination fuses dreams and reality, the eye and body, as drama. The space of imagination is there as long as subjective experience is not estranged from reason. Without *logos,* the dreamer dreams on, and in a world without *mythos* the space of imagination collapses and vanishes.

If *naming* things allows our mind to absorb the world, *ordering* things creates a new world of and for the mind, which guides and motivates our imagination and actions. In other words, when we act according to our concepts and rearrange the world as a reflection of those concepts, we discover, to our surprise, that it never turns out as we had supposed it would. If we pay attention, though, behind such disappointment lies insight—to be found in the space of imagination. Such insight precedes action; it is a creative movement of self into the world—in other words, of enactment.

BETWEEN RECORDING AND REPRESENTING

Take, for example, the story of Butades of Sicyon as told by the Roman author and natural philosopher Pliny the Elder (23–79).[5] While not necessarily historically factual, this famous description of the origin of drawing is nonetheless illustrative of how drawing may have originated with projection.

Butades' daughter Kora is heartbroken at the prospect of her lover's departure; the order of her world is about to change and disappoint. She imagines projecting his shadow onto a wall and tracing it with a stick of charcoal. The imagination here isn't some act of mimesis transposing an internally imagined image; rather the whole of it, the imagination, is born and formed outside of her mind—the trace, the light and shadow, and the image. Her act, though, is internally motivated by love, and while the accuracy of the outline is mnemonically important, the feelings and act of positioning and drawing her lover's profile is probably more memorable. In other words, the projection, while ostensibly about measure, moving data from here to there, is wholly carried by feelings and meaning—although these are hidden in the depths of the projection.

To quote French poet Stéphane Mallarmé (1842–1898), the projected shadow became a "magical shadow with symbolic powers."[6] Unlike a locket of hair or other memento, the image was magical in that it enabled Kora, indeed empowered her (one source of the English word "magic" is the Old Persian *magush* and its root *magh-*, meaning "to have power") to see and feel her absent lover in an Apollonian image—a representation of her Dionysian pain, dream fused to reality—in a work of imagination.

Projection is associated with the immaterial and is generally considered a purely propagative and invisible system of casting information. In this story, though, projection's role as an agent of Kora's love is framed in material aspects critical to the material imagination. The details are meaningful and necessary to complete the love story: the flickering light from the oil lamp; the fixing of a slightly blurry shadow of Kora's lover; the drawn qualities of the outline; her lover turning his head just so and not seeing as she touches his shadow with charcoal, their shadows merging in the act of tracing; the distancing between them done in order to make the image. As the material imagination takes hold and the formal imagination is displaced, projection reveals the hidden tensions of its drama: light/dark, fire/night, flicker/hold, ray/shadow, touch/trace, mark/smudge, stone/charcoal, burning/burnt, present/absent, warm skin/cool surface, seeing/remembering.

Butades went on to make a clay relief from the tracing, a part of the story mostly ignored by architects and painters: many of the seventeenth-, eighteenth-, and even nineteenth-century paintings depicting this story are titled the "origin of painting" or the "origin of drawing."

The relief relies less on technique than a practiced hand to build from the action of the trace. Bachelard observed that clay is to the hand as the perfect form is to the eye; he refers to it as the "imaginary paste" of the material imagination, "the perfect synthesis of resistance and malleability."[7] The making of the relief, triggered by the trace, follows its own process; only the profile trace and memory of the young man are transmitted. Here material demonstrates both its autonomy—the material directs the making of the image—and its generative capacity: the relief, fired along with roofing tiles (Butades was a potter), became first an iconic medallion, then a new architectural element known as the antefix, a decorative tile that provides a vertical termination for building roofs (figure 6.1). The fact that antefix faces are not profiles but frontal reliefs attests to the plasticity of idea: at a certain point, the original *techne* of the trace projection has been forgotten, although we can still make out the narrative affinities between the projected shadow and antefix.

The creative process is a kind of chain reaction. One thing leads to another, seemingly turning and twisting without direction. But direction *does* emerge. More precisely, what emerges is trajectory—moved by a motive force. But what is that force? Everything changes: technique, material,

6.1
Antefix, head of Medusa, Greek, southern Italy, fourth century BCE.

image. Yet something passes through, something to which we cannot quite give a name, something not even visible, but something knowable the way a person is. Everything changes—until it doesn't.

The "origin of painting," a story built on a chain reaction of insights and events, was revived in the seventeenth century by painters attracted in part to the romantic content of the subject; their various stylings lost contact with the core content of the story as they teased and tugged it into various embellished compositions. In many examples, the depictions of the projection and shadow are erroneous despite a careful attention to secondary aspects such as the draping of clothes.

Joachim von Sandrart (figure 6.2), in his 1675 illustration, draws the shadow as if it were drawn in sunlight; an accurate shadow from a nearby light source would loom massively and distort the relative size of the two standing profiles. Jean-Baptiste Regnault's 1785 painting *L'origine de la*

6.2
Die Tochter des Töpfers Butades, Joachim von Sandrart, 1675.

6.3
L'origine de la peinture, Jean-Baptiste Regnault, 1785.

peinture (figure 6.3) depicts a shadowless maiden's arm holding the charcoal. Both artists dispense with the annoying projection geometries. Never mind the mechanics; they're after the gist of the story.

Originally, the geometric precision of the projection, extromissively representing the desire to see, to remember while empirically making the image, could carry the imagination between inner feeling and material sensation. But any connection to that thread is broken by the turn to iconographically focused arrangements of the now mythological characters. The memory of the story no longer matters, only its history.

SHADOWS

In the Butades story, the shadow plays multiple roles: phenomenon, instrument, and trace. The shadow trace, a semisolidified recording of the lover's presence, is a representation as well, but the key is that the shadow trace is a fact—in time and space. Experience, generally speaking, is temporal. Geometry is spatial. Kept apart, the former is blind and the latter is empty. In this case, though, the lover's trace is both the geometry of experience and the experience of geometry and, like a good metaphor, is able to describe an abstraction—her yearning—in concrete terms.

Shadows anchor us. They are tethered to everything that touches the ground, and they detach from us when we jump. This attachment to form makes shadows feel semisolid—they are harbingers of solidity, an integral part of spatially perceived "reality." Shadows give mass, definition, corporeality, texture, and heft. Temporally, however, shadows are transient, not yet being but becoming (foreshadowing), a materialization, and a residue, a ghost, a departing ("only a shadow of himself"). Our cognition of shadows is such that we pay little attention to them. We're hardwired not to, since they are transient phenomena.

Yet shadows are instrumental. Early astronomers saw and learned more from eclipses and other forms of shadow than from direct observation. In Galileo's time, the empiricist's insistence on direct observation as the only legitimate way of knowing limited what could be learned about the cosmos, and the medievalist allowance for extraperceptual insights, such as speculative mystical conjecture, had nothing to contribute to what we would consider scientific inquiry.

Galileo's breakthroughs came in part from his understanding of how to use shadows to extend his powers of observation. At the time he trained his telescope on Venus, it was believed the planet was luminous from its own light and orbited in an epicycle—an orbit independent of the sun. Galileo saw that the planet was in partial shadow as it went through its phases, and thus had to be a tenebrous, dark body. He also realized from the logic of the shadow that Venus orbited the sun, since all phases from new to full could be observed from earth. The final demise of the Ptolemaic system came quickly thereafter, a shadow thus shedding light on the fallacious ordering of the cosmos.

Shadows elude simple classification: they are there and not there; they render solid yet are the residue of solidity; they follow and prefigure. Children's drawings are notable for many things, but one element almost always absent is shadows. And when children do draw shadows, it is almost always by tracing real ones. What does that tell us? That children aren't curious about shadows? That shadows are too difficult to cognize?[8] Or is it perhaps that we cannot *imagine* shadows in our mind the way we can imagine, say, a chair?

Although possessing geometry, a shadow has no form; no mental image can be attached to the word "shadow." It is impossible to imagine and then draw a shadow, unless you draw the form casting the shadow along with the shadow, which is spatially and cognitively challenging as well as too fluid to be pictorially assimilated into language in the way we can assimilate "cup" or "apple."

"Shadow" is a rich, even overused literary term: "beyond a shadow of a doubt"; "he was afraid of his own shadow"; "her deed will cast a long shadow"; "those people need to be brought in from the shadows." All these examples trigger associations, even quasi-pictorial ones. We've seen a long shadow, so the vague memory of it is enough to complete the mental "picture"—which I put in quotes to contrast it to an image with enough robustness to be visually present.

Try drawing "he was a shadow of his former self." The typical tug of war between seeing and thinking is absent in this case; our thoughts of "shadow" are estranged from the phenomenon of shadows—hence the difficulty of building an image of shadow through reverie or introspection of memory. Reflection, shadow's geometric cousin, also defies specificity in the mind. An optical mirror image is probably the most direct experience of reflection, yet its inherent optical and geometric qualities elude direct cognition. Eleanor Duckworth of Harvard's Education School has demonstrated this by having children "solve" the angle of incidence/angle of reflection question with a mirror and time.[9] It takes children hours to "get" that the two angles are equal around a normal—and it's no easier for adults.

PERSPECTIVE

Although the maiden Kora merely traced a shadow, in doing so she roused projection geometry, the systematic geometrical schematization of phenomena such as sight and light (including shadow and reflection) as well as the foundation of surveying, optics, and visual representation. The development of projection was a watershed: the reasoning mind could now insinuate itself into the senses, gaining enormous capacity to act on the world and shape it in its image.

The representations of idealized cities of the Renaissance were early examples, envisioned and propagated with projection geometry at the advent of its development as a means of designing. These images are notable for their noumenal as opposed to phenomenal nature (figure 6.4). A *noumenon*, as described by German philosopher Immanuel Kant (1724–1804), is an object or event that is posited, that is, that exists without perception or sense. Kant typically contrasted it with a *phenomenon*, which is anything that can be conceived by or is an object of the senses.[10] With these Renaissance representations, the mind has seemingly convinced itself of its power to create a new world out of whole cloth, without reference to physical contingency.

This wasn't the end or beginning of projection as a mode of thought, but it marked a hinge point: the medieval sense of building as an act, as

6.4

La veduta di città ideale, Francesco di Giorgio Martini (attrib.), 1477.

the outcome of an exquisite choreography of labor, materials, and techniques, began pivoting toward seeing building as an approximation of a "blueprint," itself understood as the most accurate representation of an ideal mental image.

Filippo Brunelleschi (1377–1446), the Italian designer and architect, is typically credited with the rediscovery of linear perspective. He set up his famous demonstration at the baptistery in Florence, fusing a perspective representation to the actual scene by incorporating a mirror. The viewer peered through a small opening to look at a mirror image of the baptistery drawing, superimposed on the actual view of the baptistery (figure 6.5). For the painting's sky, Brunelleschi used polished metal leaf—which reflected the real sky, further conflating the two images.

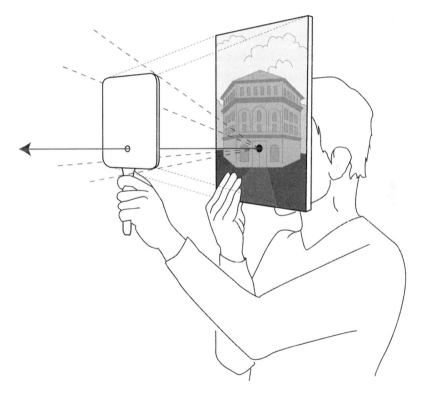

6.5

Brunelleschi's baptistery demonstration.

Like Pliny the Elder's report of the origin of painting, this "discovery of perspective" comes to us in a testament—in this case written several decades after the event by Brunelleschi's biographer Antonio Manetti (1423–1497), himself a mathematician and architect.[11] Unlike Pliny's story, Brunelleschi's demonstration never became the subject of paintings or mythologizing. It was too "modern" an event and impossible to envision, given all the awkward arrangements of viewpoints and the reversed panel and mirror. Elements of pathos or emotions are absent in Manetti's recounting, which limits the narrative appeal, as does the fact that nothing particular came of young Brunelleschi's demonstration.

According to Manetti, the baptistery panel was done systematically, not empirically, but we have no evidence other than the words of an exuberant biographer. What we do know is that Brunelleschi had earlier measured several works of Roman architecture and that he had experience with surveying, both of which rely on projection techniques. That is not much to go on. Late medieval surveying, astronomy, and proto-perspective were very close in their concepts and techniques, systematically transposing lines and angles (figure 6.6). Both surveying and astronomy relied on simple instruments which allowed for sighting angles, and in the case of surveying, leveraging of proportional triangles to make measurements of distant things, such as towers and tops of buildings.

All these endeavors moved information from field to drawing, using a common language of lines of measured angle and length. These drawings had to be orthogonal projections, specifically horizontal and vertical planes, to preserve the measurements translated from angles and proportions. Perspective relies on such drawings to be constructed accurately, and it is conceivable that Brunelleschi did some survey or measuring work of the baptistery and constructed his setup and panel from plans and sections. That would have made the task precise and systematic. Perhaps most importantly, the precise geometry of the mirror-panel-baptistery relationship would have been easy to control.

Knowing what painters already knew about lines converging to vanishing points located on a horizon line, Brunelleschi could have neatened up the work, hybridizing physical measurement and perspective projection

6.6
Etching of surveying techniques, Jacob Koebel, 1531.

principles and making it what Manetti called "scientific." If, in fact, the baptistery painting was initially intended to be viewed frontally, and given that the subject is symmetrical—a mirror image matches the image—the mirror might have been an impromptu addition after the painting was finished.

Like shadows, reflections are difficult to hold in the mind, especially their phenomenal qualities. Does this suggest Brunelleschi first made the painting and then realized that by adding the mirror the demonstration would be far more effective? The sky could now be mirrored on the panel and the panel itself became an optical image seen *through* a mirror, just as the baptistery itself was seen "through" the mirror. Intentional or not, the mirror would have enhanced the image on the panel; since mirror images are flat, robbed of their depth, the panel would appear to match reality when both were viewed as mirror images.

If Brunelleschi did work iteratively, where was he going with all this? It would seem enough to make a demonstrably "accurate" painting of the baptistery, then set it up to be viewed so as to have it match the original. The medieval populace would have been suitably impressed to see a picture match reality, especially if done with "scientific"—that is, reproducible—methods. But Brunelleschi must have realized the mirror and silver leaf would heighten the effect and at the same time be even more of a tour de force. He was notoriously paranoid about having his ideas stolen, so the obfuscatory function of the mirror would have been useful. Mirrors were themselves considered part of optics—in particular catoptrics, the science of reflections—so the inclusion of the mirror would immediately add scientific *gravitas*. The result, nonetheless, was a marriage of the senses and mind, image and picture. This "scientific" image—a Trojan horse corresponding with "objective reality"—had managed to penetrate the sacrosanct surface of painting's picture plane, which until then had been the exclusive purview of subjective representations of the imagination.

It is difficult to know what impact Brunelleschi's panel had on the development of systematic perspective. Certainly there's little if any evidence of his later influence: the historical narratives that have sprung mushroom-like from Manetti's words seem all to seek confirmation that Brunelleschi "invented" perspective right there in front of the baptistery. I would argue that Brunelleschi did something less: he simply applied known measuring techniques in widespread use by surveyors of the time to known perspective rules. But in doing so, he produced a representation that claimed for itself an unprecedented degree of "truth" by relying solely on technique, whereas typically it was the subject matter—mostly religious—that had previously carried this authority. Brunelleschi made perspective itself the subject of his painting—the mirror view of the actual panel pushing it, literally, further into and conforming to the background. The implications for perspective in painting weren't realized for some time, a lack of progeny not helping matters.

Brunelleschi moved on to architecture. An examination of his designs yields clues to his methods of representation and demonstrates the influence of drawing tools on his architectural thinking. The Pazzi Chapel in

Florence, an early project, departs radically from the Romanesque construction sensibility wherein space and order are the result of geometries of structural organization and construction techniques that are governed in turn by the characteristics of materials: weight, strength, the structural limits of spans, vaults, buttresses, and the like. Brunelleschi is credited with bringing autonomous abstract geometric orders to architecture, accompanying a revival of classical ordering principles. In the Pazzi Chapel, abstract immaterial geometric order is foregrounded while the Romanesque constructional aspects recede and even disappear behind the new geometry. The classical orders, ostensibly rooted in structure, are here overlaid in a superficial manner, with no correspondence to actual structure. An inside corner of the chapel "swallows" its pilasters to maintain conformity with the rules of the geometric layout (figure 6.7). The odd result is visually "wrong"—but correct if generated in plan and elevation drawings laid out with uninflected geometries.

It is ironic that Brunelleschi, the engineer, would be the first to reconceive physical structure as a purely visual problem (the classical orders) with barely any regard for the actual structural questions. Of interest here is the essential role projection had in this new way of thinking about a building. Projection drawing is necessary in order to draw abstract geometry. Yet drawing first and building from drawing yields an inevitable clash between the immaterial abstractions afforded in drawn architectural plans and the physical and material logics of construction. Brunelleschi used projection drawing to create a design for the Pazzi Chapel, and the building has acquired characteristics of projection: flatness, abstraction, and planar geometry. It is a testament to the influence of tools on thought and intent, even while architects believed their design ideas sprang freely from the mind, and subsequent buildings shared these tendencies throughout the Renaissance.

This bifurcation of the subject of structure into geometric absolutes on the one hand and contingencies of construction (now hidden) on the other has echoes of the baptistery panel. There too, geometry—the geometry of perspective—is triumphant, while the business of its construction is left out of the story. The likelihood of Brunelleschi having used plan and

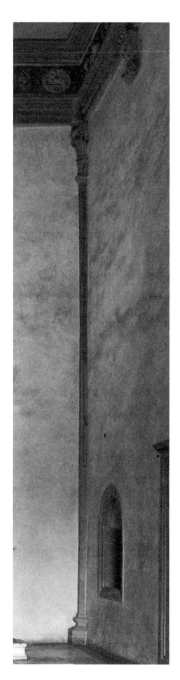

6.7
Swallowed pilaster, Pazzi Chapel, Florence, Italy, Filippo Brunelleschi, 1443.

section drawings (horizontal and vertical measured drawings, respectively) as part of his perspective method as well as his building design is not absolute, although it is almost certain for the latter. What does seem consistent is his willingness to mix empirical knowledge with theoretical propositions.

While Brunelleschi moved on, perspective continued to develop two intertwined conceptual paths, *perspectiva artificialis* and *perspectiva naturalis*. The former concerned itself with drawing outward—that is, the representation of space using geometric perspective—whereas the latter was about drawing in, using perspective principles to record an existing spatial arrangement onto the page or eye. Medieval optics, which included the study of mirrors and reflections, constituted much of *perspectiva naturalis*, but as Brunelleschi's experiment exemplified, projection theory and techniques were the subject matter of both *artificialis* and *naturalis*. His drawing of the baptistery belongs to both, having the ambition to be a representation *and* a recording of reality.

The early Renaissance saw the development of perspective in treatises and practice. Decades after Brunelleschi's experiment, the Italian polymath Leon Battista Alberti (1404–1472) published his treatise on perspective with no reference to Brunelleschi's baptistery "invention."[12] He laid out a system that is today the basis for computational modeling—namely, a conception of space as a three-dimensional, gridded, measurable volume.

Painting, which long before Brunelleschi had employed vanishing points, horizon lines, and measuring lines (the setting out of points and lines that determines a perspective drawing), did not incorporate highly systematic perspective until the *Holy Trinity* (c. 1427) by the Italian painter Masaccio (Tommaso di Ser Giovanni di Simone; 1401–1428), which was painted at least twenty years after the baptistery demonstration. Renaissance painters after Masaccio were quick to adopt Alberti's system; notwithstanding the system's secular/scientific trajectory, they incorporated projection geometry into their religious subject matter. In the painting space, perspective was recognized as "God's geometry" and was deployed with all sorts of inventiveness: a "womb axis" in which vanishing points (the point where receding parallel lines, such as train tracks, appear to converge in a perspective) coincide with swollen bellies; light rays emitted by annunciating doves; divisions of netherworld and heaven made to coincide with point of view and horizon lines; and other marriages of piety and rationality.

Art historians have drawn distinctions between northern and southern European painting, particularly Florence's development of projection geometry, *perspectiva artificialis*, and the northern painters' attention to optical effects such as reflections and luster, more the purview of *perspectiva naturalis*. While artists continued to embrace the mysteries and narrative content of religious painting, the techniques and geometries of projection slowly wrested "truth" from its metaphysical wrappings, replacing mystery with metrics and displacing the symbol-rich allegorical landscape of painting with something more pictorial. Painting ceded its emotional depth to perspectival and illusionistic depth. Put another way, *wonder*, that opening of self into depth, was replaced by *marvel* and *spectacle*, which operate at an emotional and physical distance. Further developments in projection, while reinforcing this tendency, also generated rich new territory for the imagination to inhabit.

The Italian painter Piero della Francesca (c. 1415–1492), in his treatise *De prospectiva pingendi*, worked through the unresolved aspects of Alberti's system, in particular the perspectival treatment of complex objects including the human figure.[13] Piero's "other method" worked from the object to the picture plane, while Alberti's system of working from the gridded field could handle things easily only if they were geometrically related to the field, such as rectilinear blocks and rooms. Piero's approach utilized auxiliary drawings, plans, and sections to get to the perspective; the shocking horizontal "slicing" of the human head (as a necessary part of the measuring process), while only incidental to the final perspective, is deeply objectifying (figure 6.8).

Piero used Albertian perspective in his paintings to order space and buildings, while his figures, having been conceived by the "other method," float serenely and independently. He had the ability to blend the systems to achieve pictorial unity, but chose not to. His paintings are paradoxical—a strong sense of abstraction pervades the figures—but rather than distancing, the effect is like that of numbers, so interior to us that we cannot see

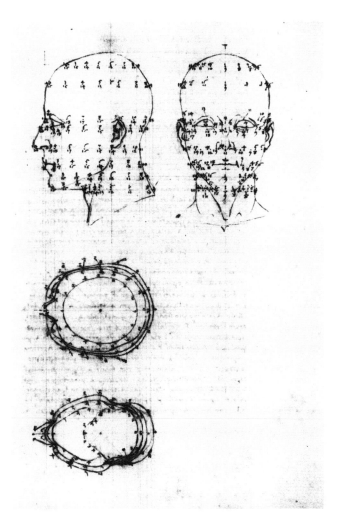

6.8
Elevations and section cuts of a human head, Piero della Francesca, c. 1474.

them as "other" or "out there." The Albertian context in Piero's paintings is more fixed and visually stable. The contrast between the two systems is strange yet familiar—a slice through life's subject/object contradictions.

In Germany, the contributions of painter and printmaker Albrecht Dürer (1471–1528) to perspective—of the *perspectiva naturalis* sort—were more instrumental. His interest was to capture existing objects accurately with mechanical devices that tracked rays to a viewpoint, intersected with a frame holding a grid or paper. Of his three published perspective drawing arrangements, the most jarring has a reclining nude being "objectively" drawn with the aid of grids and a sighting point (figure 6.9). It is difficult to ignore the implications of the omniscient eye, the voyeuristic gaze, and the objectification of the female body; measuring in this case has given up any claims to innocence. As a perspective apparatus, it is rudimentary, more aid than instrument, since the draftsman has to "eyeball" between what he views on the vertical grid and what he draws on the grid's twin on the table. In spite of this, however, both draftsman and subject are frozen, lashed to the system in the service of measure, in contrast to the independence of painters and their muses to move freely, to roam, to look from different points, to accumulate views.

Dürer's most sophisticated mechanism (figure 6.10), involving string, a hinged frame, and two operators, is remarkable for what is missing: there is no artist or viewer present or necessary, the problem of perspective having been reduced to a series of discrete tasks involving nothing more than marking. The two fellows in the print pointedly are not attending to the image but to their operations. Conceptually, the arrangement is a machine, anticipating the photographic revolution by three centuries.

Unlike Piero's work, Dürer's perspective apparatus seems to have had little if any influence on his paintings, in which he closely studied optical phenomena. Historian Jan Białostocki has noticed how Dürer paid a lot

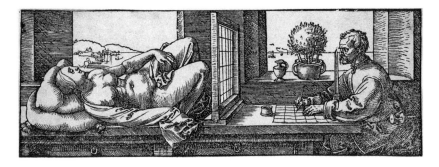

6.9
Drawing apparatus and nude, "Der Zeichner des liegenden Weibes," Albrecht Dürer, 1512.

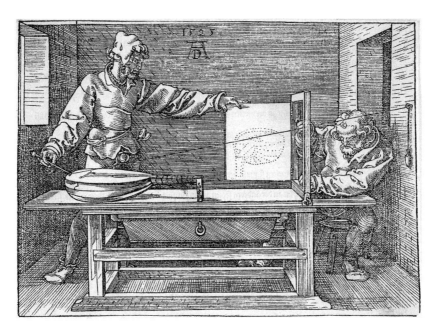

6.10
Hinged-door drawing apparatus, "Stich aus Dürers Anweisung zur Messung mit Zirkel und Richtscheid," Albrecht Dürer, 1525.

of attention to the reflections in eyes of his subject.[14] This attention didn't arise from his work on perspective devices, which only reproduced form measurements; rather, it is from the northern European practice of depicting convex mirror reflections in paintings and thus extending space outside of the frame, *The Arnolfini Portrait* by Flemish painter Jan van Eyck (before c. 1390–1441) being the best-known example (figure 6.11).

Eyeballs have distinct convex curvatures, different from the whites that hold them, and Dürer understood these could only be distinguished by the reflections of light sources on the shiny surfaces. As his painting progressed, he began rendering eyes with precise reflections, over time changing from indistinguishable highlights to clearly visible divided lites of a window (in architecture, a lite is a term for a piece of glass, and in windows, the term refers to separately framed pieces or panes of glass). Was this an astute observation by the painter in his studio? It would appear so, especially if

6.11
Detail from *The Arnolfini Portrait*, Jan van Eyck, 1434.

the major light source was from such a window. His fellow German paint-ers subsequently painted windows in their eye reflections as well, appropri-ating his insightful perception as an applicable technique.

CONFLICTS BETWEEN PERCEPTION AND PERSPECTIVE

When a perception is grasped by the mind, seeing is both powered and blinkered in a single stroke. Once painters could "see" the window reflec-tion in every painted eye, they painted it. They painted it regardless of the optics, outdoor scenes now sporting impossible window reflections in eyes and shiny tankards. Even Dürer's exquisite painting of a hare in the snow features a window reflection (figure 6.12). Some have argued this is evidence the hare was painted in studio, and others that Dürer used the window reflection as a painterly device without reference to its context.

Regardless, Dürer the observer and Dürer the painter were at this point self-aware; vision is more than any of the other senses a cooperative project

6.12
Detail from *Feldhase*, Albrecht Dürer, 1502.

of thought and perception. Cooperation doesn't mean balance or symmetry; as the hare example shows, thought has a way of structuring things and projecting those structures forward in time, whereas raw vision, if it exists, must disorder, must forget, in order to open itself to the present, to the fullness of phenomenal perception, to see "for the first time." That "first time" moment is, paradoxically, the wonderful experience of grasping, of *aaahhaa*, when something sensed is mentally perceived—made conscious, just before being seized by thought, not unlike the way a good metaphor can establish a relationship where one did not previously exist. Once the first time gives way to a second and third time, recognition absorbs it, thinks, knows it, and fixes it—in mind, ready to deploy, turned away from receptivity. But seeing isn't a matter of eyeball aiming: the hand that draws a reflection deepens vision markedly and creates cognizance. At the same time, when something is absorbed, when it is "seen" with action (especially the hand's action), it tacitly embeds in procedural memory, whose architecture is only configured for generalities that can be reenacted, the antithesis it would seem of an optically precise and unique event.

The act of drawing is similar to working with material: there is a lot of back and forth between initial intent, resistance, reflection, realization, and action. For anyone who draws, especially students of drawing, this process is a familiar one that repeats itself over and over. First there is always a struggle to see and draw, the two informing and confounding each other, the drawn line rarely being "exactly what I wanted." This felt dissatisfaction, which compels to action, eventually—through effort—leads to insight and then to procedural mastery. This is followed by deftness, confidence, and technical repetitiveness—and then a plateau, a sense of staleness, of being "in a rut" (of procedural consistency), which marks the beginning of the next cycle.

Insights, like metaphors, don't disappear but are assimilated into quotidian life and culture. The window reflection on eye is now found on every shiny sphere in the Sunday comics, having evolved from a technique to a graphic shorthand, denoted instead of connoted. Meaning is absent. However, in Dürer's time it was meaning that made optical reflections such an enduring feature of painting. Eyes were understood to be powerful things and so were reflections, since the space of these reflections was charged, lying outside our field of vision, seen only obliquely, suggesting an analogue of the relation between secular and sacred realms. At that time, eyes—being the proverbial "window to the soul"—were imbued with religious potential; the magical relationship of the three elements popularized by Dürer—reflections, eyes, and windows—was bound to be recognized for its symbolic content. Sure enough, the combination of window, eye, and reflection appeared in religious paintings; more tellingly, the actual windowpanes in the reflection started being rendered asymmetrically so that the divided lites resembled a cross.[15] Like the devalued use of the reflection meme, the crucifix/window frame lost its optical truth with repeated use.

Signs and symbols stand between images and concepts, but operate somewhat differently. Signs are arbitrary and transparent, like "1" standing for "one." Symbols are opaque, rich, and multivalent. The window, by transforming into a literal symbol (the cross), lost its transparency and, in the process, its signlike aspect—the inevitable end product of an elaborate chain of insights and concepts. Beginning with the original and precise recording of an optical condition (reflection), the phenomenon had slowly transformed into something more narrative and iconographic in nature. Seeing was swallowed by meaning.

The iconic eye reflections of Dürer and his colleagues would seem to foreshadow the enduring value of meaning in the field of painting. After all, if meaning is making sense of life, the role of painting would not be in question, at least to the fifteenth- and sixteenth-century mind. As *perspectiva naturalis* and *perspectiva artificialis* continued to develop, their shared techniques and intents began to unwind, as the question of truth shifted from the metaphysical ground to a scientific or objective framework. For *perspectiva naturalis*, truth became instrumental truth, with the development of tools and devices to record images, ranging from the *camera obscura* (the reversed and inverted "pinhole image" projected through a small hole onto a surface opposite the opening) to the *camera lucida* (an optical device that superimposes an image onto a drawing surface) and eventually to film. Even without the use of devices (although Vermeer's work is possibly an exception, as one contemporary inventor speculates in a 2013 documentary film),[16] painting continued to develop by interrogating the qualitative aspects of visual phenomena—that is, the way things *seem to appear*. Painters from J. M. W. Turner to the pre-Raphaelites to the impressionists rendered such immeasurable phenomena as weather, the glint of sunlight, and reflections in rippled water as precise visual sensations. Meanwhile, meaning—at least iconographic meaning—retreated from being the eye's mind.

Perspectiva naturalis concerns itself with what exists—in other words, what is always in the past even at the moment it is "captured." *Perspectiva artificialis* concerns itself with the future, with what is not yet, which is why it has been of particular importance to architects. The question of truth for *perspectiva artificialis* manifested itself in greater and greater control over metaphysical ambiguities such as shadows (sciography). Ultimately, advances in projection swept away the transcendent space of fifteenth-century Italian artists' perspective. One of them, the sculptor Donatello (Donato di Niccolò di Betto Bardi; c. 1386–1466), went so far as to include elaborate checkerboard floors in the simple mangers that were part of his portrayals of Christ's birth—checkerboards that were obviously nowhere to be found in the original manger, but that adhered to the primary laws of geometrical perspective. But eventually, even the qualia of the mind's eye—the imagination—found projection, and particularly perspective, to be a constraining system.

Only a few later artists, such as the great Italian printmaker Giovanni Battista Piranesi (1720–1778), continued to use perspective as an instrument of the imagination: finding refuge in the oblique, thus mirroring the eye's physiology, in which the peripheral is ruled by the subconscious and emotional aspects of vision. Piranesi's famous *Carceri d'invenzione* (imaginary prisons) are a tour de force exploration in and of perspective, each of the 16 etchings constructed as a diagonal view, both horizontally and vertically. "The Gothic Arch" (figure 6.13) is an example. The cues that typically provide control and measure—ground, horizon, a set of vanishing points—are missing or hidden. Rather than a cool and controlled perspectival space, the *Carceri* present a dizzying, disorienting, and exhilarating emotional journey into what feels like the imagination itself.

The influence of perspective on the history of European painting is well understood. Its deeper effect on our (culturally contingent) self-awareness is less obvious. Notwithstanding a hundred years of nonrepresentational art, perspective is still regarded as *the* default visual representational system. The history of perspective is seen as a progressive and triumphant unfolding of greater and greater geometric control over phenomena. By that measure, medieval and Byzantine perspective is considered primitive, naive,

6.13

"The Gothic Arch," Giovanni Battista Piranesi, c. 1749–1750.

and wrong, out of touch with "objective reality." Take for instance, the 1405 *Annunciation* (figure 6.14) by Russian icon painter Andrei Rublev (1360s–c. 1428), which is typical in its use of "reverse perspective." Parallel lines converge forward, which gives an effect (to our linear-perspective-trained eyes) of elements getting larger as they recede. There is also the matter of the building on the upper left, which seems to be showing three sides at once and thus offers no sign of a systemic unity in the geometry of the lines. Each element is depicted independently of the others, as if we are looking at fragments of experience.

In his 1894 lecture titled "Why Has Man Two Eyes?" physicist and philosopher Ernst Mach (1838–1916) addressed the matter of perspective.

It seems to you a trifling matter to look at a picture and understand its perspective. Yet centuries elapsed before humanity came fully to appreciate this trifle, and even the majority of you first learned it from education. I can remember very distinctly that at three years of age all perspective drawings appeared to me as gross caricatures of objects. I could not understand why artists made tables so broad at one end and so narrow at the other. Real tables seemed to me just as broad at one end as at the other, because my eye made and interpreted its calculations without my intervention. But that the picture of the table on the plane surface was not to be conceived as a plane painted surface but stood for a table and so was to be imaged with all the attributes of extension was a joke that I did not understand. But I have the consolation that whole nations have not understood it.[17]

6.14
Annunciation, Andrei Rublev, 1405.

Mach reminds us that perspective, rather than representing reality, deeply conflicts with our own perceptions and being-in-the-world. In reality, vision is a scan over time, never a frozen image, and as we *move*, a three-sided view of a building is closer to experience than a one-sided or two-sided one. A scene before us has no "perspective" as all elements and edges endlessly shift their perceived geometry. In what is known as *parallax*—the visual phenomenon stemming from one's movement through space—elements appear to shift relative to each other at varying rates depending on their distance from the viewer. With a flick of the mind, we can focus on and fix the background and witness the foreground move. Convergence is a momentary and local condition, every instant changing. Depth, flatness, and binocularity come in and out of our visual field depending on what we are looking at.

Perspective is difficult for children to comprehend. Its mastery is not an intuitive extension of perception; rather, its geometric schema must be learned and practiced like any new language. It's not in our nature. In other words, the physiology of our vision, when we examine it from experience, has nothing to do with perspective or, for that matter, with any geometrically unified system that defines space.

And what is space? You're unlikely to get the same answer from one person to the next, because there is no singular agreement about it. Space can be defined geometrically, physically, or perceptually, and likely all three at once, without any conceptual connection between them—although perspective, it is claimed, relates geometry to our sense of vision.[18] Considering the multiple realities of space, the schema of perspective forces us to dispense with a lot, although for the most part we hardly notice. When we do—such as when drawing from life in contrast to drawing from a flat image—the difference, which is vast, is treated more as a "degree of difficulty" technical challenge. No matter what we are drawing, when following the schema of perspective or another projection system, we abandon all other sensory information that doesn't "fit"—but as we have seen, that means abandoning our eyes' direct contact with reality. It also means that the world is reduced to skin, surface, since drawing from a single viewpoint cannot penetrate form, and the depicted space is obliged to be homogeneous, lattice-like, and infinite, despite spatial experience being anything but that.[19]

Mach went further in his "Contributions to the Analysis of Sensation":

> A common and popular way of thinking and speaking is to contrast "appearance" with "reality." A pencil held in front of us in the air is seen by us as straight; dip it into the water, and we see it crooked. In the latter case we say that the pencil appears crooked, but is in reality straight. But what justifies us in declaring one fact rather than another to be the reality, and degrading the other to the level of appearance?[20]

Our era's answer to Mach's question is *objectivity*, which defines the discipline of science. However, our humanity depends on sensitivity and the validity of experience. Visual artists are charged with seeing, seeing more, penetrating, and empathizing with the mysteries of reality; to treat "appearance" as less factual than "reality" is blinkering. Perspective, the objectification of representation, projects a similar bias—anything that doesn't fit the system is no longer a fact of reality.

Returning to Andrei Rublev's painting (figure 6.14): if its "appearances" do not line up with any single unified system, it may mean that the artist was more in contact with his perceptions and imagination than another artist who faithfully executed a technical perspective. But vision, which we claim as our own, is culturally situated, and Rublev's painting is one of countless works in the Byzantine tradition that share similar traits. Like the cross in the eyes, experience has become codified and signified to stand between perception and concept. That Rublev's painting is symbolic doesn't detract from it being richly rooted in a world in which "appearance" is reality.

That leaves the reverse perspective as a head-scratching trait. It's difficult to imagine any circumstances in which objects grow larger as they recede. It has been observed that if you imagine looking down at a surface depicted in reverse perspective but then imagine that it is actually overhead, the surface will appear to be in "normal" perspective, diminishing as it recedes in the pictorial space. Contemporary Russian artist Gor Chahal (b. 1961) has gone one step further and drawn a striking connection between linear and reverse perspective (figure 6.15). In turns out that if a perspective is extended past the vanishing point, an inverse and reversed perspective image is realized, as turning figure 6.15 upside down reveals.

In other words, if the Byzantine painters wished to represent spiritual space as beyond the horizon, on the other side of infinity, then reverse perspective was a brilliant, revolutionary insight, and one that would tend to be adopted as a representational strategy—which it was. Though reverse perspective was a recognized trait, Byzantine artists resisted making a geometrically unified (and confining) system out of reverse perspective. Why would they need to, if the spaces they represented were otherworldly and metaphysical?

Most images are easily "consumed"—consider pictures we simply "look at." For an image to resist such simple consumption, it has to trigger something in us such that we take some action: engage with, recreate, and complete the image, not unlike the way a great metaphor is experienced. The Byzantine picture plane's empathic correspondence with the viewer involves fragments of vivid experience and imagination to be retraced and reexperienced, like a dream memory, in a flowing medium unrestricted by

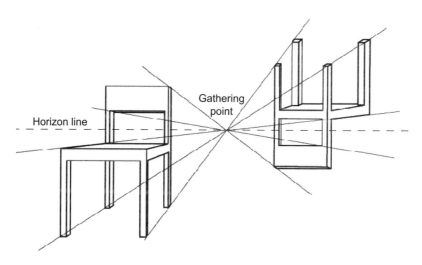

6.15
Reverse perspective using a chair, Gor Chahal, 2010.

rules of projection. This recalls the reunification sought in Greek tragic theater, similar to the Apollonian construction in the spectator's mind of the Dionysian oneness with the world.

The origins of linear perspective lay in the making of illusions, dating back to Aristotle's reference to the introduction of *skenographia*—set painting—by Sophocles and then its more widespread use by later Greek playwrights.[21] They likely understood perspectival set painting, using it to create illusionistic space on a surface for the spectator to take in as part of the spectacle. As Greek theater evolved from the original raw tragic form, it was lightened and transformed into what we might today consider a play in the modern sense: something to enjoy at a distance, as part of an audience. Scenographic perspective replaced reality with its outward image, consistent and paralleling the transformation of Greek tragic form from trembling lived experience into entertainment and acting.

Projection and perspective drawing are taught as technique, the emphasis being on the *how*. The unnaturalness of it makes projection drawing somewhat like flying blind, with a heavy reliance on instruments and procedures. When projection is conceived as something technical, it remains insistently rigid, demanding adherence to method and rules aimed at maintaining the "naturalness" of perspective, to avoid revealing the glaring incongruities with actual perception. Those that know it only as technical drawing are never going to achieve anything besides, well, a technical drawing.

The artists of the Renaissance were able to transcend this limitation, conceiving of projection, and perspective in particular, as a language—an indeterminate system to be used, to be manipulated as a means to make an image. The works of the great Italian polymath Leonardo da Vinci (1452–1519) and Piero della Francesca are "wrong" from a technical standpoint, but it is their "mistakes" that transport us into the work, past the screen of superficial resemblance.

7 THE TRANSMITTING HAND

Drawing plays a role in making, whether of stone carving, toaster, apparel, or building. Each discipline involved in making has developed techniques to move between the two-dimensional and three-dimensional, the most simple and direct of which—templates and profiles—are magical insomuch that they metamorphose from drawings into physical tools that are then directly involved in the process. Besides their function of transferring metrical information from one place to another, all tools and techniques involved in 2D/3D translation have an outsized impact on whatever they touch. Still, while ostensibly designed to aid the process of making, these tools and techniques influence and limit it as well.

DRAWING AND MAKING

Drawing lies at the intersection of the fine arts and design, promising to be the universal medium, bonding artists and designers in a common language. Everyone sketches, uses paper and pencil, draws representations, makes marks for their own sake, and draws things that don't exist but will. However, there are differences between disciplines in the way drawing is used to translate between two and three (or even four) dimensions and how such translations are seen to affect the authenticity of a work.

The Renaissance was borne on the development of orthographic drawing techniques. We can observe the effects in painting, with the use of orthographic projection to systemize perspective, and in architecture, with the increasing frontality and flattening of surfaces toward relief, which corresponds to the plane of the orthogonal drawing from which a building has been translated. Sculpture was less influenced by the new methods of projection, since it didn't rely on a translation from drawing as in the case of architecture.

Historically, sculpture has held to a romantic myth of direct engagement by artist and chisel, putting aside the whole business of tools and tricks developed to aid in creating the work. And for good reason: the sculptor's authorship was a signature, bound to the object itself. There could be no distance between the work and the hand that created it. Indeed, this Dionysian unity of matter and artist stands in stark contrast to the unerring execution of a plan. For the artist there is no plan; the material and maker ecstatically give themselves to each other, and the emergent work is thus born as "art."

Jack Rich expressed a disdain for "indirect carving" (carving with the aid of measuring tools) in his 1947 book *The Material and Methods of Sculpture*. He wrote, "It is definitely difficult to retain vitality in a work when it is reproduced mechanically from clay or plaster into stone or wood, particularly since most indirect sculptors have a meager knowledge, and experience with, wood and stone."[1] While Rich regarded mechanical techniques of reproduction as devitalizing, he also acknowledged that "direct carving" was not without its critics:

> Some substantial criticism has been directed against the sculptor who respects his medium to the extent of desiring a minimum of wasted stone, and who adopts his concept to imperfections that are uncovered within the block as the work progresses. The critics maintain that the rough shape of the stone as it comes to the sculptor is accidental at best, and that he is therefore handicapped, since the medium should be controlled by the sculptor, rather than the reverse.[2]

According to this critique, material must submit to method. Material is inconsequential to the intention of the artist, and whatever its flaws, they must not be allowed to interfere. Others, as Rich notes, have argued that material itself is too intimidating and "infuses the carving with a fear of failure that hampers the hand and fetters the imagination."[3] For the carver, the balance of hand and material is found in a "Goldilocks" zone in which the material does not overly challenge or unduly influence the sculptor's concept—it's "just right." Even the great Italian artist and architect Michelangelo (1475–1564), the exemplar of direct carving, used models and tools to find balance in the carving process.

Rich drew a line, though, at mechanical devices, claiming their use led to devitalizing and uncreative outcomes—what I might even call *slick*. One type of device, the pointing machine, is used to copy a model made in a (usually additive) material to a block of another material, allowing it to be carved accurately. After taking point measurements with a rod held in a frame, the entire rig can then be moved to the new block, and the measurement translated by drilling holes into the block until the rod fits exactly. Pointing is an early analog version of multiaxis CNC (computer numerically controlled) devices that bypass the flattening tendency of orthographic projection now so entwined in architectural thinking and production. The irony is that because of pointing's great accuracy, it is considered deadening and undermining of the hand.

The chassis or frame method, similar to multiaxis CNC but conceptually related to orthographic projection, is less criticized because it is clumsier (and therefore less dominant over the hand). The measurements must be taken in the x, y, z axes of the frame consisting of four reference planes. If a surface of a model is at an acute angle relative to the axis of measurement (such as measuring the side of a nose with a stick held perpendicular to the frontal plane of the face), accuracy is degraded, which puts the onus back on the hand, as one supposes, to bring the work to artistic completion. Other mechanical aids such as calipers, profile gauges, and templates escape inauthenticity because they are tools directed by the hands—they are extensions, not replacements, of artistic skill.

An architect carving a sculpture would probably work in a fashion similar to Michelangelo. He was unique in his planar approach to sculpture, often drawing on the face of a block (an orthographic projection) and carving along the projection axis, working into the block from front to back. He would dip a wax model of his work into water to see the animation of the form as it emerged through the surface, subtracted from solid, released from captivity. The contours of the semisubmerged model would be valuable. Such greatly simplified lines are easily transferable by "eye" or instrument. They also describe planar slices through form. This is essentially an architectural strategy for translation between 2D and 3D; in this instance, the model's form is analyzed and reduced to a stack of contoured planes, each of which is then sequentially reconstituted as a guide for the carving as it proceeds from the front to back of the block, first contour to last.

It is commonly remarked that Michelangelo's genius lay in being able to see the finished work inside the solid block and, by free and direct carving, "releasing" it from the mass. No doubt drawing and dipping models in water helped him see, but what about the frontal approach to carving? He could have dipped his models from all sides, but chose to work only from one.

Even before the advent of perspective, painting and drawing tended to reflect a consistent preconception, namely, that space opens away from and behind the surface of the canvas or paper. Representations lie at various distances back in this spatial matrix—which perspective systemized but left conceptually unaltered. Michelangelo's drawings and paintings invert this concept. His surfaces mark the boundary of solid and space, solidity extending infinitely behind.

Historian of art and architecture Alina Payne has observed of Michelangelo's *Madonna and Child* that the image emerges from the plane of the page as if being excavated (figure 7.1).[4] Michelangelo's *Last Judgment* is also rather shallow; its figures tend to be close to the surface, almost relief-like in their composition. Nowhere are there indications of deep space.

In some of Michelangelo's architectural studies, the design elements are drawn in elevation as a series of overlapping and shifting outlines, suggesting a stacked and compressed view of a body's many contours. It implies that he carved as he drew and drew as he carved, turning "translation" (2D/3D) from a technical issue to one of artistic process, one involving the direct and tacit experience of making, in particular carving. Imposing a limit to one's freedom of action, as with Michelangelo's frontal, drawing-bound approach to sculpture, may seem inhibiting. Why do it?

7.1
Madonna and Child, Michelangelo, 1533–1534.

While drawing occupies a territory between mind and artifact, it is neither, and the imagination can either daydream or push against physical reality, such as with the "excavation drawings." If the drawing harbors the seed of the form, if it is more blueprint than first sketch, its projection will then be a persistent and authoritative guide. It has also been speculated that the frontal approach allowed a greater degree of freedom over time, since the rear of the block remained open to revision even as the front was the site of activity.[5] This is analogous to tracing a line over a line, changing it with each searching trace and slowly bringing form out of the surface.

It is one thing to understand Michelangelo's process in the context of his paintings and sculpture; it is another to consider his architectural work, which like the work of all architects involves a relegation of authorship to the craftsmen building the work. Michelangelo (and the other architects of his time) designed projects using wooden models and drawings (his with a subtractive and layered bent), but in some cases, such as the Medici chapel at San Lorenzo, the work is a fusion of sculpture and architecture, of Michelangelo and craftsmen.

Michelangelo's architectural work is distinctive for the freedom with which he stretched, pulled, and bulged the classical conventions, ignoring ordering and proportioning geometries. His drawings are similarly without an underlying geometric framework to restrict boundaries. But if such "soft" drawings are the point of conception of the project, how can this be transposed into something definitive without losing the sensuous qualities suggested by the blurry tracery of the drawing? The answer lies in the drawings Michelangelo developed to guide the fabrication of his architectural work.

Modani are full-scale profile templates, first drawn and then cut from paper. The craftsman transfers these into rigid material to be used as instruments to guide the carving of the profiles. How can cutting a hard edge of a profile drawing preserve the soft ambiguity of multiple line tracings? It is in the development of the *modani* themselves.

The architectural historian Jonathan Foote has found that Michelangelo's *modani* seem to be drawn without instruments, bearing no marks of a compass or ruler.[6] He documents Michelangelo's process: first he drew on paper, then made adjustments by tracing over it, using various media (red chalk, black chalk, ink) for various purposes, each act differentiated. He then cut out the negative, proceeding to use the template to trace the verso, "stretching" and "pulling" the tracery by shifting and sliding the template, transforming the profile like a palimpsest. When the craftsman received the final profile, it had history, it had been worked, and this could not have gone unnoticed. It would have affected the craftsman's sensibility: the edge of the profile was not some machine-made ideal to mimic and follow, but a "living" line with which to work (figure 7.2).

The craftsman is like a tightrope walker who works in exquisite tension between eyes, body, gravity, and the line of the rope, only in this case the quivering rope was Michelangelo's *modano*. This calls for empathy from the

7.2
A *modano* for a cornice, the Medici Chapel, Michelangelo (undated).

craftsman: empathy for the material, yes, but more critically for the artist via his template.

Empathy is not passive. It is activated with give and take, with a conversation. Such reciprocity doesn't happen between a craftsman and a conventional guideline. These *modani* were the result of give and take, pushing and pulling, which in turn invited Michelangelo's craftsmen to join in, to add their voices as they "spoke" with the stone.

Up to this point, we have traced how the hand can remain present in a work—even if it doesn't quite come in contact with it. Michelangelo's *modani* aren't like architectural blueprints; rather, they have retained and passed on the touch of their maker. Do the *modani* mark a limit beyond which making is conceived as indirect, even as anonymous, with no connection to the hand? For the fine arts, this marks a threshold, but for designers and architects it is met with a shrug. To the latter, there are other ways drawing can transmit "the hand."

A teacher of mine, the painter Paul Rotterdam, taught a valuable lesson on the first day of class. We were all told to draw horizontal lines on a sheet of newsprint mounted on the wall. The only "rule" was to focus and carefully follow the previous line by drawing just below it. This proved easier said than done.

As line after line (and there were hundreds) was drawn, each was subjected to his scrutinizing eye. Like a cardiologist examining a chart, he would point out our shortcomings here and there. "You are thinking too much," he would say, or "you are thinking too much ahead" or "you have stopped paying attention" or "you are trying too hard." It was tiring, and, as lines accumulated, it became more difficult to find the balance point between effort and indifference.

The "authenticity" of the hand, if defined as that point of effortless focus, is sporadic; the ability to act appropriately in response, to listen to what the eye sees and the hand feels, has to be cultivated. When we speak of "the hand" it is shorthand for such intimate reciprocity, or physical empathy. Just as empathy can be extended across space to others at a distance, so too can "the hand"—as long as this back-and-forth between feeling and acting is unbroken.

How does such empathic sensibility find its way through the lengthy design process? Álvaro Siza, a contemporary Portuguese architect well known for his cafe napkin drawings, made a squiggly line, just so, to describe the roof profile of a Berlin building he was beginning to design. The actual building facade turns the corner with an unusual languid roof parapet profile, appearing as if drawn with *that* pen on *that* napkin, though not necessarily resembling it. We know the napkin sketch wasn't reproduced and enlarged by some template, or by dedicated assistants, and we know there were many intermediate drawings and modeling used to slowly draw the line into form which had to incorporate the economic, technical, zoning, and other forces at play. In this case, "the hand" was found within an empathic process of development envisioned, not ruled, by a serendipitous moment in a cafe.

St. Peter's Church in Klippan, Sweden, designed in 1963 by the Swedish architect Sigurd Lewerentz (1885–1975), is an anomaly, its blueprints purposely eclipsed by construction events and discoveries. The architect was continuously *on site* working with the situation *at hand* and made design decisions in the moment. This could have been chaotic had there not been sufficient guideposts. The blueprints were only for a kind of overview and remained very general, so some other means of understanding the relationship between a detail and the overall was required. Lewerentz used the brick as a "ruler" by forbidding any brick to be cut. All measurements had to be in full brick dimensions. This constraint gave the masons and Lewerentz a newfound creative freedom, since they were working the brick as both the design and construction medium. Lewerentz's "hand" was found in the intense reciprocity between the act of building and the design reactions (figure 7.3).

LE CORBUSIER AND RONCHAMP

For Charles-Édouard Jeanneret (1887–1965), the Swiss-French architect known as Le Corbusier, "the hand" was a powerful symbol. He described his Open Hand Monument in Chandigarh, India, and his many other Open Hand sculptures as visual acoustic sculptures, representing attentiveness

7.3
Baptistery, Church at Klippan, Sweden, Sigurd Lewerentz, 1963.

and the capacity to give and to receive.[7] He spoke of his design of the chapel of Notre-Dame du Haut in Ronchamp, France (figures 7.4 and 7.5), completed in 1954, as having sprung from a long subconscious period of gestation. With charcoal in hand, he was able to "give birth onto the paper."[8] The drawing itself reminds me of the fluid motion associated with forming clay on a potter's wheel.

The building itself, however, was controlled by strict geometries that exactly dictated the outcome. How, then, can "the hand" claim its place in the process?

The initial sketch (figure 7.4) consists of four concave gestures—like open hands facing outward. The finished building is not hand-shaped or shaped by hand, yet more than any building of its time it resists association with being machine-made or guided by rules of geometry. The ceiling/roof in particular expresses a corporeal sagging that evokes the tremendous weight and deformations of clay. Metaphorically and metaphysically, "the hand" and its empathic function and value had all entered the design process.

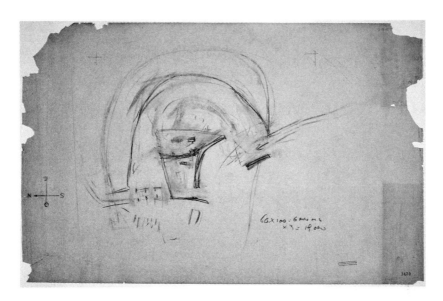

7.4
Early plan of Ronchamp chapel, Le Corbusier, undated.

7.5
Ronchamp exterior, Le Corbusier, 1954.

Reinforcing the mystical narrative, Le Corbusier said he found inspiration from an empty crab shell he happened upon at a Long Island beach. The knowledge of it came through "the hand." He kept it as part of his collection of poetic objects, where it sat on his desk until he was inspired to make it the roof of Ronchamp.

Words have the power to rearrange our world, and the shell became roof-shell. The shell's form and exoskeletal strength, but also the metaphorical association—the void left behind by a spent life—may have made the shell seem right. From the start of Le Corbusier's design process, "the hand" had several roles in the conception and translation of the roof: the oneiric hand gave birth onto the paper and the physical hand touched and translated one shell into another, with the design of the roof still to come.

Published study models of Ronchamp are made of string, wire, and tissue paper, quite unclaylike. If there were any preceding clay models, they do not figure into this tale. Molding clay would not confer authorship unless it is near its final scale, as in the case of sculpture. Simply translating a crab shape would do little, the roof ending up as nothing more than a bad copy of a crab shell. Architecture must constantly leave behind the materiality of its contingent form in order to build deeper ties with its eventual construction logic; otherwise, it will be limited to hollow emulation. Like a good metaphor, architecture thrives when it can be two very different things at once. In this case, the geometric studies in wire and string turned out to be a precise logic for construction *and* the means to carry the narrative of the primal gesture.

The secret of Ronchamp was in its execution. The strict geometries were internalized and are nowhere to be seen. In the ceiling, the sagging, almost limp mass is accentuated by the curving outlines of the boards used to form the concrete (figure 7.6). This is camouflage. The roof geometry is constructed of straight lines, as demonstrated by the string and wire models. Why not, then, lay the formwork along those lines? It would have been much simpler to keep the boards flat, but it also would have revealed the controlling line geometry of the ceiling.

The ceiling, despite its appearance, is developed from a conoid, one of the simplest geometric solids to draw. The conoid is constructed from

7.6
Ronchamp interior, Le Corbusier, 1954.

a circle, a rectangle, and a triangle in the three orthogonal *x*, *y*, *z* views. It belongs to the family of ruled surfaces—those surfaces that can be described by straight lines. But if only a segment of one is used, it can make a surface appear as corporeal and enigmatic as the one at Ronchamp.

Michelangelo's design for the entry to the Laurentian Library is celebrated for its fluidity and daring stretching of the classical orders. Despite his limited involvement, having sketched a design that he described in a letter to an associate as having come to him in a dream, Michelangelo is accorded a status of direct authorship. Associates used a model apparently made by him (and since lost) to guide the fabrication of the finished staircase and entry. It should be noted that the model—being made from clay, soft, easily pressed and stretched—points to material's generative, or at least participatory, role in the stair design. Michelangelo's hand was thus entangled in the unique and unrepeatable interactions of correspondences, material, craftsmen, and associates.

The Laurentian Library entry is a more valued and "untouchable" work than that of any architect, in part because the architect's work must pass through the anonymity and reproducibility of construction documents. Once there are such documents, the architect is dispensable. Le Corbusier knew this dispensability lay ahead when he made the first sketch of the chapel—that the most raw and subjective mark of "the hand" would have to pass through the design and drawing process to emerge as a rational, measurable set of instructions. Had he ignored that and continued the design process according to the terms of his original conception, the geometries and construction process would have been imposed from without as technical methods, yielding a result not dissimilar to the indirect methods of sculpting that Jack Rich found so devitalizing.

Michelangelo transmitted the vitality in his hand while still employing indirect methods. Both he and Le Corbusier understood that a work's development through drawing and/or design process could not be separated from its means of execution. For both artists, the techniques and methods of translating the sketch, the model, the block, and the found were part of the continuum of creativity and, importantly, gave the work the vitality we recognize as authorship. Compared to Michelangelo's San Lorenzo, Le Corbusier at Ronchamp had to go through additional steps, transforming the work from a mythopoetic narrative (the dreamt gesture and the crab shell) to an analogous geometry and purely metrical set of instructions in order to construct it.

The working drawings of the roof of Ronchamp are gossamer-like, akin to that of an aircraft wing, yet the roof itself seems leaden and distorted by gravity's pull. Robin Evans noted this was, in effect, the inverse of the Gothic cathedral—which strove to render mass weightless, in part by highlighting the geometric tracery of its vaulting.[9] The chapel plan can be understood similarly: the convex walls and outdoor pulpit are consistent with a church turned inside out.

Le Corbusier produced his painting of an "Apollo/Medusa" hybrid image in 1945, marking a turgid period for the architect disillusioned by World War II's destructive triumph of technology. For Le Corbusier, the hope for the future was in laying bare the truth of our primal roots, exposed through a kind of alchemical reaction between antagonists, a "unity of opposites" even if that unity was a violent clash. At Ronchamp, the opposites are primal gesturing and coolly rational geometry, which fuse together and contradict each other, one presented in the form of the other, Apollo covered by the mask of Medusa.

Paul Klee described Art as the breathing of life into the inanimate,[10] the sending of the metaphysical arrow just a tiny bit further than what is physically possible—"the hand" being regarded as a hallowed instrument and transmitting agent for the artist. This presumption makes intermediary drawing (drawing in order to make) suspect; any intervening medium can only dull, lessen, dumb down, or at best propagate. If, however, we allow for the possibility that drawing itself has agency, we may begin to see it as a catalyst, a transformer, even a point of origin of artistic insight. Ronchamp's richness of associations, multiple conversations with history, and density of content are due in great part to the continuous and lengthy metamorphic chain of drawings, from initial sketches to precise working drawings, and not despite them.

EMPATHY: FEELING AT A DISTANCE

I have discussed the role of the hand as a transmitter of vitality, of feeling, of empathy between material and artist, between architect and building, via drawing and other means. In the case of Michelangelo, the transmission of the profile through drawing, cutting, and carving involved increasing degrees of resistance extending from the drawn line to rigid *modani* and finally to the stone itself, which in its material stubbornness transmitted back to the craftsman and the *modani*. How could Le Corbusier (or any designer) working at a distance from the constructed artifact "feel" its material resistance? The key is in the way media were used—not as representational tools, but as engaging materials—in the sense that their limitations and affordances were consciously allowed to drive the thinking process as much as they were driven by it.

As Jerzy Soltan, an architect and former employee of Le Corbusier's atelier, reminisced:

> At the rue de Sèvres [Le Corbusier's studio], the model did not have to be precise or particularly elegant to fulfill its purpose. . . . What charcoal represents on the plane of a drawing, plasticine represents in volume. Little bricks of this

fat clay are not only easily cut but are easily malleable. I watched them under Corbu's fingers. I saw the charcoal sketches begin to appear as interpretations of the clumsy, topsy-turvy clay toys, illustrating the burgeoning idea of, perhaps, the city center of Saint-Dié or the first proposal of the housing development at La Rochelle. The charcoal followed the spatial study in clay. The charcoal sketch was translated once again into the language of plasticine and then went back on paper. And so it went until the black ink of the fountain pen fixed the project for a while.[11]

Note the reciprocity between the two ways of working and the two materials charcoal and clay. Soft and malleable, they are close enough in their qualities to engage in dialogue, while sufficiently apart in their respective realms (drawing and modeling) and materiality to resist each other.

Le Corbusier's Open Hand Monument in Chandigarh was preceded by his *Poem of the Right Angle*, a series of 19 paintings and corresponding writings issued between 1947 and 1953 and arranged in seven rows or "zones." The text for the sixth, *Offre* ("Offering—The Open Hand"), reads as follows:

> It is open because
> all is present available
> knowable
> Open to receive
> Open also that others
> might come and take
> The waters flow
> the sun provides light
> Complexities have woven
> their fabric
> the fluids are everywhere.
> Tools in the hand
> Caresses from the hand
> Life is tasted through
> the kneading of hands
> eyesight resides in

palpation
————————

Full hand I received
full hand I now give[12]

The final lines are emphatic: the hand transmits in both directions; it receives and it gives. A small detail of the sculpture is also telling. The colossal metal hand is mounted on a swiveling shaft that allows it to turn in the wind. Forces guide the hand. The confluence of giving and receiving, acting and reacting, speaking and listening all characterize a larger dimension of empathy and the participatory capacity of material that arises from its own reciprocity and resistance.

At this point, it may be useful to take stock of what is meant by "empathy." *Einfühlung*—German for "feeling into," as empathy was originally defined—allows us to extend our emotive being to others, to feel and understand their interior selves and their perspectives of mind. Empathy can be described as the felt immediacy that binds us. We think of empathy as direct—a person feeling another's pain, or sadness, or some other emotion—without the intervention of any agent. Empathy as a feeling arises as a recognition of the familiar, and so the strange and unfamiliar are most difficult to empathize with. This innate human capacity contributes to the resilience of society, equips us to accept differences with others, to navigate change with less fractiousness. Such a propensity to identify with people derives from our senses and extends to things in which we delight and with which we identify and which we speak of as beautiful.

All this raises a question: Is empathy no more than the innate ability to receive through the senses, or does it include a sort of probing, outward mental "tool" that can be developed in humans? Whatever the answer, it is a form of drawing near—becoming "closer" either by tuning one's "receptivity" or by projecting oneself. The receiving and giving hand is the empathetic hand.

The connection between material and empathy is most evident as a close-in give and take, whether between people or between a craftsperson and material. Like electricity, empathy flows with contact and the close presence of the objects of our attention. This form of empathy comes naturally—a pressing of the elbow, a handshake are both part of a haptic form of empathy. An example of its measurability is evidenced in a well-known 2008 study by psychologists that demonstrated how touching a warm cup of coffee led to increased empathy, the physical warmth of the cup leading to mental feelings of warmth toward others.[13]

It is much more difficult to feel for others or for material when at a distance, be it physical distance or the mental distance that accompanies forms of thought such as theory and scientific reasoning, especially when these are decoupled from tacit or embodied knowledge. It was for this reason that the English poet, painter, and printmaker William Blake (1757–1827) famously criticized his fellow Englishman, the mathematician, astronomer, and physicist Isaac Newton (1642–1727), accusing him of "single vision" disconnected from the sensibility and spirit of being.[14] Blake considered Newton to be denying God by describing the universe mechanically and mechanistically. As he wrote in a letter: "Pray God us keep / From Single vision & Newton's sleep."[15] Blake also lampooned Newton in a caricature, depicting him as senseless, naked, and cavebound while engrossed in the figures and formulations of his Newtonian reality (figure 7.7).

Yet empathy extends beyond the proximate and physical and sensed. We've constructed a world of representations and projections of ourselves, our fears and doubts, within the "real world" of things and phenomena, forces and matter. Deploying all our possessions, clothing, built environments, devices, symbols, and language, we actively instrumentalize this relationship with the world. We've learned to feel, to empathize through projections and representations: a picture of a loved one evokes feelings as real as the original. Even reading about someone's travails can move us enough to make a donation.

In chapter 3, I wrote about affective and cognitive empathy as mental processes, terms used in social psychology to describe direct emotional and distanced thoughtful forms of empathy, respectively. Haptic empathy, which is related to affective empathy, refers less to the mental processes and more to the physical context that transmits empathic feelings. What, then, are the vehicles and mechanisms of transmission that extend empathy

7.7
Newton, William Blake, 1795.

beyond the direct and local to include cognitive empathy (or, as it is also referred to, perspective taking)?

It seems something similar to how the Apollonian and Dionysian urges were brought together in the birth of Greek tragedy is at work in empathy when active at a distance. In Greek theater, there is a chain of transmission from the writhing mass of revelers to its revivification by the ecstatic, music-infused, dithyrambic chorus to the actor, pivotal as both a hallucination of the chorus and a representation of it to the audience. This representation allows the audience access to the Dionysian feeling through Apollonian contemplation. This chain is not free of friction or resistance and, in my opinion, the effort required along each link of the transmission is what generates the electricity of empathy—similar to how physical and material resistance conduct haptic empathy.

Given that feeling shapes how a design process is steered by designers and experienced by others once completed, empathy could be characterized as a designer's most important sensibility. If design is to be an empathic process, how can empathy exist between designers, their designs, and their audience—especially when the employment of the distancing instruments of projection and scale and time in the design process closes off haptic empathy? As with Greek tragedy, there has to be a chain of transmission.

A key transmitter can be found in surrogates—the so-called "empathy belly" that simulates one aspect of pregnancy is a well-known example that allows one to empathize with the experience of another. In design, materials are almost always used as surrogates for other materials or even for formal concepts. Car modelers, as they shape clay, understand they are working vicariously with a substitute material that has nothing to do with sheet metal, yet they feel an empathic connection with the sheet metal through their work. As Damian Lottner, a clay modeler at Ford Motor Company's design studio in Cologne, Germany, has explained, "When we as clay modelers work on the [full-scale model of the] car with our hands, we also transport emotions."[16]

Architects also rely on surrogates. We build models and make drawings as we probe a design, bringing reasoning and feeling to bear. Visual, spatial, and physical phenomena can be modeled, studied, and guided, giving access to acts of inhabitation and even to a vicarious sense of just how the light will feel coming through a particular opening. Materials also play a mnemonic role in the design process: for instance, linseed oil on a smooth, thick piece of walnut can flood one with memories and feelings as thoroughly as a madeleine did for Marcel Proust.[17] In all these examples, empathy is still transmitted with feeling, sensing, and thinking hands. The more this happens, the more a work emerges with its materiality integral to the design.

The links in the empathic chain of transmission remain distinct and close enough to each other to be intelligible. In the example of Ronchamp, each step along the design path had to hand off to the next step, so that a crab shell could be a transmitter of emotion into the conoid geometry of the roof. That the roof geometry has nothing to do with the geometry of the crab shell indicates that what is being transmitted is neither measurable nor rational but emotional. The example of Ronchamp applies to every design process: the empathic chain of transmission depends initially on the hand, but for it to continue beyond such a limit the developing process must bring its own narrative into play. As with theater, design is the creation of a story, and it is the story that transmits empathy; the experience of Ronchamp *feels* different when its mythological, oneiric origins are part of that experience.

I've tried to describe the chain of transmission as a way of creating proximity, a drawing near or making familiar enough for empathy to happen. It should be added that this cannot be forced; in fact, it depends on receptiveness, something at which the senses are far superior to thinking and volition. This is the work of untethering the self, releasing it toward the other. It is akin to what German philosopher Martin Heidegger (1889–1976) referred to as *Gelassenheit*, literally meaning "serenity" or "composure" or "poise" but which he meant as a sort of "releasement" (as it is typically translated), the giving up of willfulness, a letting go in order to be caught up in, allowing things to be whatever they may be even if we are uncertain.[18]

Efforts to induce empathy as a method have invariably yielded the opposite. Like metaphors, empathy fades with repetition and the slightest contrivance. Consider the question "How are you today?" What was once a greeting that implied genuine interest and caring is today largely used, and heard, as a social nicety—and like all related social niceties, it encodes the impulse to draw oneself near. To complicate things further, empathy has historically been suspect. After the Second World War in particular, it was considered a powerful sense that occluded the intellect and dampened the ability to make dispassionate, and thus good, judgments. As the renowned British architect James Stirling (1926–1992) wrote about Ronchamp in *Architectural Review*:

> The sensational impact of the chapel on the visitor is significantly not sustained for any great length of time and when the emotions subside there is little to appeal to the intellect, and nothing to analyze or stimulate curiosity. This entirely visual appeal and the lack of intellectual participation demanded from the public may partly account for its easy acceptance by the local population.[19]

Stirling criticized the newly completed Ronchamp chapel as being too "easy" to experience, implying that architecture had a duty to resist affective empathy and challenge passive acceptance. The project of modernism was, at its core, a refutation of subjectivity and its attendant emotional impulses. The First World War had demonstrated the dangers of unchecked passion and nonobjective inclinations, which led to the *neue Sachlichkeit* (new objectivity) of the 1920s and 1930s in Germany—a movement in art that was a reaction against expressionism. The rise of Nazism and the Second World War ended this movement, but the postwar period saw a return to a deep suspicion of propaganda, easily swayed masses, and unmediated experience.

It was in this context that Stirling wrote about Ronchamp and expressed his unease while acknowledging the local populace's embrace of the chapel. Why isn't that "easy acceptance" a sufficient outcome for a work of architecture or, for that matter, a work of art?

Artifacts of art, design, and architecture have value because they provoke, question, and challenge assumptions (even if they earn the general public's dislike in doing so). The foundations of a democratic society depend on the freedom to choose one's point of view—and Stirling faintly damns Ronchamp for its totalizing effect, which for him suppresses the "freedom" of an inquiring mind. (It should be noted that Stirling—who considered "rational" only those geometries he could see—most likely was unfamiliar with the *hidden* geometries and orders of Ronchamp. A greater understanding on Stirling's part would no doubt have mitigated his critique.)

The *episches Theater* (epic theater) movement, of which German playwright Bertolt Brecht (1898–1956) was the most famous practitioner, was developed as a critique of the passive, empathic connection between audience and actors in popular theatrical performances of his time—today better known as "tear-jerker," mawkish, or "feel-good" productions. Their effectiveness relies on a strategy of coordinating all the elements so seamlessly that the audience doesn't notice the lighting, swelling violins, background, and script working together to elicit a powerful emotional upwelling. Brecht purposefully separated the elements of production to create a distancing effect on the audience, breaking habitual (i.e., lazy) empathic connections between audience and actors. His plays would have actors step out of roles; time would be chopped into unnatural blocks; and commentary would interrupt action. The idea was to stimulate the audience's attention and concentration and have audience members construct their own meaningful insights into the characters—not only the characters' feelings but also other dimensions of their being. Epic theater practitioners called this the *Verfremdungseffekt* (estrangement effect).

Such a stimulus of estrangement isn't unemotional; it is disturbing and brings feeling and reasoning into dialogue. For Brecht, this form of critical empathy rejected the idea of empathy as exclusively an act of mindless immersion.

Brecht was working in the theater, but the interpretation of actual events presents similar conundrums. Professional historians intent on maintaining their objectivity have dismissed empathy and, in doing so, have limited the inclusion of the memories and experiences of witnesses, victims, and perpetrators in historic perspective. To bring experience more fully into history, historian Dominick LaCapra has proposed what he calls "empathic unsettlement": "And it involves virtual, not vicarious experience—that is to say, experience in which one puts oneself in the other's position without taking the place of—or speaking for—the other or becoming a surrogate victim who appropriates the victim's voice or suffering."[20]

LaCapra sees tolerance, respect for others, and respect for and acknowledgment of the differences between subjects' feelings as all part of "empathic unsettlement." For him, empathy is not limited to feeling the feelings of someone else, but is a way to construct a broader understanding of another being or, as he puts it, "a respect for the otherness of the other."[21]

"Empathic unsettlement" or critical empathy requires a conscious balancing of feeling with thinking, coupling logic with subjectivity, and in that sense is essential for any design process, which must actively navigate between the two. In the earlier examples of Michelangelo and Le Corbusier, the vast difference in scale of their respective work shifts the predominance from haptic empathy to critical empathy.

Our relationship with materials parallels the way the nature of empathy changes with distance and "unsettlement" or degree of resistance.

Touching and handling materials, especially appealing and familiar ones, fosters the development of tacit understanding, giving a sense of the limits, the fragility, the nature of material. How often has that favorite shirt been the favorite shirt because of the way it feels?

This sort of material rapport or empathy is uncritical and unconscious, part of the torrent of constant exchanges between our psyche and our environment. Despite an almost universal but vague acceptance that our environment affects our mental wellbeing and thoughts, this understanding remains difficult to articulate in language. Simply *stating* that wood is "warm" and steel is "cold" falls well short of insight—because of a lack of unsettlement.

We experience an example of material unsettlement in the ubiquitous little Dixie Cup dentists offer us for rinsing. Unsettlement doesn't come from that cup, which is made with waxed paper and raised ridges, but from an exact copy available in the finest bone porcelain, equally translucent. When picking up the porcelain version, the material shift is so contradictory that it unsettles. It's not a copy but an alchemical transformation, paper to porcelain, disposable to heirloom. Although a trivial example, the porcelain Dixie Cup gives some clues to how seamless our mind's material engagement is and how difficult it is to induce unsettlement.

How does material engage us, unsettle us, when it's at a distance? I am focusing here on the *experience* of material and not the physical working of it, something I take up in the next chapter. The discussion of Ronchamp touched on this; the "total" experience Stirling criticizes was not unsettling at all to him. The materials of the chapel, shotcrete (a type of sprayed-on cement) and cast concrete, are familiar enough to normalize the otherwise rather astonishing (even unsettling) architectural elements, such as the sagging corporeal roof and the south wall that thickens to 10 feet—quite enormous for such a small structure. Material unsettlement at such a scale is not found in surprises of substituting one material for another, but in the behavior of material over time. After all, there is nothing particularly unsettling about seeing stone replaced by wood, a substitution that informs so much of traditional New England architecture.

Weathering, a word that describes material behavior over time, is a constant and universal condition; its ubiquitous effects surround us. It is through weathering that we take the measure of material. The way it weathers, rots, deteriorates, and patinates rouses us. Though we feel an intimate familiarity with weathering, it—like a shadow—rarely fails to penetrate the imagination or to surprise with its unexpected outcomes. Time's effects on materials are often unsettling. A warped piece of wood, a rusted post, crumbling concrete, mildew-stained surfaces, each of these common examples disturbs and disrupts when it appears unexpectedly. As our buildings are increasingly constructed with a certain material timelessness in mind, the shock of seeing works 30 or 40 or 50 years after their pristine debut is often unsettling, especially if the original materials were never imagined as transforming and weathering over time.

This phenomenon presents designers and architects with one of our most difficult challenges: to imagine and empathize with material and anticipate its nature over time. Construction details are both a technical skill and an empathetic one. They demand a heightened degree of attention, more so when trying to anticipate effects decades in the future. Builders, having witnessed every construction misstep, acquire a native wisdom and inherent caution that underscores the difficulty of imagining material behavior through the instruments of design.

In the context of the mind, empathy constitutes part of the original social network that encompasses our mostly unconscious interactions with each other and our environment. But just as with the internet, connectivity comes at a cost: the seamlessness of the emotional plane of consciousness has its weaknesses. Passive acceptance is subject to manipulation and lessens the ability to differentiate. It is self-reinforcing and tends toward prejudice, fixity, preconceptions, and habituation. Seamlessness, moreover, makes it difficult to grasp where empathy ends and intuition or imagination or outside influence begins.

Designers and architects are called upon continually to empathize with the inhabitant or the client. At the same time, that responsibility encompasses a larger sphere. Often clients come to us with a list of wants that need to be understood and respected. But our more important responsibility is to unearth their essential needs. To do so, the narrative arc or story clients tell themselves must be challenged, and alternative narrative constructions offered. Breaking such a narrative arc can be perturbing to the client. It raises questions and a need to resolve differences (or at least to hold them together). It also opens the opportunity to retell, to develop a new narrative inclusive of the contradictory crosscurrents.

This is where imagination and empathy intersect—in the narrative construction of new meaning from disparate parts. Both imagination and empathy as combinatory impulses engage the mind's capacity to bring contradictory elements and materials into a new relationship.

More commonly, this is called *creativity*.

8 WORKING WITH MATERIAL

I've discussed the relation of drawing and making in the context of the singular, the one-off. With mass production—repetitive making—materials tend to throw intention off track, which becomes *the* problem. It is why every society has developed methods and skills to control materials, to make them do what we intend. Craftsmen, their guilds, and apprenticeships have all been part of the project to master materials—and, of course, so has industry. With craft, the creativity of the artist is subordinated to the need for uniform production, although the bond between maker and product—a product's authorship, the touch and mark of the hand—remains relevant; in contrast, industry seeks anonymity and uniformity. We might value the handmade in certain instances and not in others. "Hand-carved" would not be a suitable attribute for a piece of machinery, but seems to be desirable for our sliced ham.

CREATIVITY VERSUS PRODUCTIVITY

Production and creativity are antithetical; any creative process suspends intent and expectations of outcome, which are core determinants of any production method. Design, industrial design in particular, straddles the two realms, but there the creative process doesn't emulate production. The values of production—methodology, predictability, repeatability, uniformity, and efficiency—are exactly what any creative process must jettison in order to proceed.

Artists and designers expect to be confounded. They expect to be drawn into the unknown, especially at the uncertain early stages of a process. How different this is from those who work with methods to control an outcome. A material-based creative process begins with an action, and it is through such action that materials reveal their capacity to surprise us. These surprises come in many varieties. For instance, an architecture student of mine tried to find an organizing principle generated from a specific controlled action and chance outcomes. She began with the action of clamping multiple sheets of window glass together and gradually tightening the clamp until the glass broke, which it did in a way that could not have been predicted. Creativity needs to proceed with some kind of resistance to or derailing of intent, presumptions, and preconceptions. In short, creativity must undo that which is known. After all, to be creative is to bring something new into being, not propagate what is already known.

Materials reveal problems and flaws, which are the very things production and craft work very hard to eliminate. For example, the first lenses were highly flawed, full of bubbles that interfered with optics. Subsequent production techniques aimed to eliminate these bubbles. While production techniques have been innovative and brilliant at finding ways to eliminate flaws, the creative process is just as likely to use flaws as agents of resistance or affordance, a kind of springboard to leap to other actions, to launch a conversation with and through material, leading to what previously could not have been imagined. The American furniture maker George Nakashima (1905–1990) is famous for working with boards other makers might consider "imperfect" because of their deep cracks. In fact, those cracks inspired his celebrated butterfly joints, which ultimately led to his signature tables that contrast the raw (and flawed) with the precision and control of those joints.[1]

ATTENTION

Mastery of a medium or technique can lead to an instrumental relationship in which tools (or instruments or machines) dominate the exchange between author and work. In a conversational relationship, tools are more subordinated to the author, such as when a painter uses a paintbrush. How, then, can we master tacit knowledge and, with that knowledge, gain a degree of command while remaining open and free within the medium being worked? I think of the woodworker who has deep understanding, who can look at an oak board and know just how *that* tree grew and which side of the trunk was in the sunlight, and use that knowledge to select the board to make the table that will be straight and warp-free. All that knowhow is goal-oriented and masterful: the woodworker is working technologically—which etymologically can be understood as *making logically*. The logic of this kind of mastery is like that of good chess players, who have such a depth of knowledge that they understand the interdependence of actions and anticipate consequences far ahead.

The challenge for anyone working creatively is to use one's mastery to raise new questions and "disregard the rules," as the Renaissance artists demonstrated with perspective. For example, contemporary Italian sculptor Giuseppe Penone has taken large individual timbers and, by carving away everything newer than a certain annular ring, "unearthed" the young tree within, branches included (whose remnants are the knots of the timber, as in figure 8.1). He has had the sensitivity *and* freedom to allow the timber to guide him to original work.[2]

In retrospect, it may look easy, but when facing a stubborn material or medium, one that resists our intentions, it is natural to feel a greater degree of difficulty and awkwardness and longing for mastery of the *how*. "If only I knew how to . . . !" The appeal of achieving a desired result has its place, but if creativity is the making of *something* from *nothing* or bringing the new into being, then a goal or result that does not exist cannot be known and thus cannot be aimed for. So how do we proceed? Questions, knowledge, observations, suppositions, and hunches all contribute to framing the situation, to grounding it, regardless of whether an artist is starting a painting or a designer is challenged to design a piece of furniture. The first encounter with any physical medium—pencil dragging on paper, a lump of clay squeezed, a chunk of wood hefted in the hand—starts us on a journey. Thoughts and feelings that come to mind are already entangled with the tacit, embodied knowledge we carry in our bodies, and in the material

8.1
Tree of 12 Metres 1980–2, Giuseppe Penone.

at hand. This is so natural that it is difficult to notice: we are prone to think our thoughts spring only from our minds.

The initial stirring to action, before anything has happened, seems to evoke what is commonly mischaracterized as our *imagination*. It's a feeling, a burst of confident anticipation that finds its form in words: "I imagine . . ." or "I'm going to . . ." or the less insistent "I'm interested in . . ."

What these declarations share is a strong sense of being unfettered. All is possible within one's self, one's "I": "I imagine," "I'm going to," "I'm interested" are all voices of intention, a turning toward, facing the situation, paying attention, focusing energy.

What happens next is action. Hands move; tools are assembled and deployed; media and matter are rubbed, pressed, cut, molded, folded, strung, brushed. In short, all these things are *acted on*. At this point, training, experience, and skill will carry and propel the work only to a certain point. Knowhow has a trajectory and wants to follow it; however, any creative process demands openness to the unexpected, a responsiveness to the situation at hand, a recognition that what we "have in mind" may not be at all what is beginning to transpire before us.

Every process unfolds in its own unique way through bumps and failures. Forward motion and backtracking are familiar features and essential aspects of creativity. Why? Materials and media have *resistance*. They resist our efforts to force them to conform to intent. They wreck our plans. They can become the boulder blocking the road. Over the years, I've watched hundreds of students arrive at that moment of frustration and boredom, ready to give up. It is truly a remarkable moment: the emotional tenor seems so pessimistic, yet it is a necessary precondition for true insights. Resistance wakens the imagination.

All materials and physical media resist and react to action on them, each in its own profound way. Creativity is birthed in the crucible of resistance. When habitual ways are stymied, when we can no longer proceed, creativity becomes *necessary*. But first, before you think "what should I do?," attention must shift to what is actually happening *out there*, to feeling it.

To empathize, to *feel into*, is the most intimate way of knowing another, including a material other. The American architect Louis Kahn (1901–1974) expressed this idea when he spoke of asking a brick what it wants to be: "What do you want, brick?"[3] Asking such a question and listening to the answer requires action. The kind of action is less critical than the material's response—how you strike a bell is less important than listening to the sound it makes. No medium, material, or technique is knowable until it is queried with action.

I realized this as a graduate student trying to make a work with two materials, lead and copper. I had tried for weeks to hammer the lead into sheets thin enough to be moved by ambient airflow when made into a mobile, using copper wire for the armature. The results were hopeless: the lead sheets could never be hammered thin enough to lose their clunkiness, and the mobiles I made were absolutely immobile. This dead end turned out to be a new beginning when I came across a geological testing laboratory on campus with an incredibly powerful press. Winding copper wire around some lead bullets (figure 8.2), I could *sense* the different attributes of the two materials that weren't visually apparent. And then the press did a magical thing. Under the enormous pressure the copper wire and lead expressed their traits, the lead "flowing" bound by the "stretching" copper (figure 8.3). Ordinarily hidden, their respective materiality appeared under the action of the press. From there, the work was propelled.

Material, as that experience illustrates, doesn't give up insights easily. As much as forceful action reveals, it is initially and in most regards a pretty blunt act, hemmed in by preconceptions of how to act. Eventually, though, *something* gives—this is where mind and material begin to converse. More

8.2
Wound copper around lead bullets, Christopher Bardt.

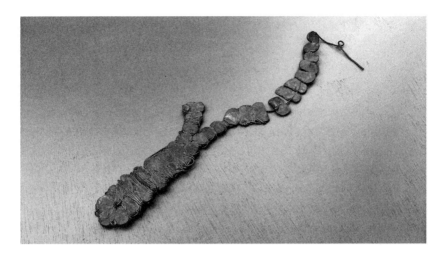

8.3
Wound copper and bullets expressing their relative traits after being pressed, Christopher Bardt.

often than not, that initial conversation is easily derailed: a satisfying result tends to beget more of the same action that led to that result. This is where intent must exhaust itself and eventually let go.

The process of letting go is challenging. It lies beyond reason. It is guided by feeling, by emotional sensitivity and sensibility—our most vulnerable antennae. With time and practice, we may learn to understand eagerness, annoyance, intuition, and digging in our heels as critical guideposts along the creative path, as necessary preparations for arriving at a state of disinterest. As actions beget reactions and interactions, and as intent is confronted with the autonomous behavior of material, "trying too hard" and "want" are natural instincts to match stubbornness (of material) with more of our own stubbornness. New possibilities always drift into view when desire is exhausted—marked by a state of not wanting, Heidegger's *Gelassenheit*.

Eventually, the resistance of materials opens the space of reverie, that half-dreaming state in which the imagination can stir and where dispassionate contemplation penetrates far deeper than any overt effort.

MATERIAL EMPATHY AND MATERIAL IMAGINATION

Handling material pushes sensation to the fore, particularly touch and the sense of weight, heft, fragility, and the feel of things. Muscles and reflexes register forces and absorb them as tacit knowledge; working with physical reality shapes muscle memory, craft skill, physical skill, and athleticism. This type of knowledge is rather impervious to thought and therefore to language. It's very difficult to "think" of how to throw a ceramic pot— you've got to do it, take action, if you are to feel it. In his book *The Inner Game of Tennis*, W. Timothy Gallwey makes the point that only when the inner voice is silent, allowing the unconscious or nonverbal self to perform, is it possible for there to be a unity of being between body and ball. "You bum, you'll never learn how to hit a backhand," he writes, to describe the futility of trying to improve your swing (or any athletic action) with some scolding and instructing inner voice.[4]

There is always an exchange with worked material. It is first an exchange of action and reaction, but ultimately it becomes an exchange that binds material and author. Each of us has developed within ourselves, at some level, material empathy, the feeling for things, for the nature of material. The ability to know a medium's nature comes from engaging it, analogous to a conversation between two strangers: the more strange the other, the more work there is to overcome distance. The work itself produces feelings, sensations, emotions—an understanding that is sensory and embodied. I can't tell you much about a certain kind of paper, for example, except by working with it—and even then I can't explain what I tacitly know.

Material is slow. It is heavy, clumsy, and can be felt, unlike the gravity-free optical nimbleness of "seeing into" an image. Material is deep. It stirs primal sensibility. Playing with material penetrates our wholeness, engages our emotions, feelings, muscular wisdom, and haptic knowledge. When Gaston Bachelard wrote "matter is dreamed and not perceived," he was refuting the accepted idea that images of the imagination (perceived image) are built out of memories and those of perception out of terrestrial material.[5] Bachelard proposed that matter gives rise to a primal oneiric imagery—like that from dreams and dreaming—that evokes the imagination (imagined image). In Bachelard's sense, material leads us to images that precede and guide perception, that distort and metamorphose reality.

Philosophers and poets are inclined to consider the material imagination as something welling upward to and through language, since language is their medium of expression. Artists make images from material and are never at a remove from that material—they are never outside the realities with which they work. I imagine, for example, a graphite drawing of "graphiteness" will endlessly circle back on itself. And the quotidian nature of material, surrounding us to such a degree that we see it as "normal," gives few if any clues to its effects on our imagination.

Yet examples of simple triggering experiences with material abound. Curling our toes into warm, sunny sand sends us somewhere. I'm not talking of sweet nostalgia, but the drift and dreams that emerge when we are immersed this way. Rubbing a piece of smooth pear wood imbues us, as does feeling the warmth of a glowing ember. Handling a button made of polished horn gives pause, while a plastic one leaves us indifferent. How can these experiences with material not affect our subconscious? Touch isn't exactly action; rubbing a piece of wood isn't the same as working it, but if I touch that piece of polished pear wood without having worked it, I can know much about it empathically—and thus access material imagination.

But what is physical material anyway? We can answer this question from the specificity of our personal experiences. So, to a photographer, material is an optical phenomenon. A mason's senses concern themselves with physical properties such as a material's weight and hardness. To the passerby, material barely registers, and then mostly on a subconscious level, *unless something disrupts the norm.* Perhaps with that in mind, the legendary architectural historian and critic Colin Rowe (1920–1999) wrote that "surely, architecture always involves an element of theatrical distortion or exaggeration."[6]

Consider a brick wall in isolation and the same wall when sunlight and shadow fall on it. The former is mainly perceived optically; the latter gains the haptic—gravity, light, heat, and cold are *felt.* Surface gains texture and reveals hints of the history of its making and context, the makings of a narrative. Natural phenomena are a kind of "action" on material, revealing attributes and adding experiential, enveloping depth. Even though walking by this wall on a beautiful spring day may not elicit a conscious sensation, we are very much immersed and affected, immersion being the fullness of sense—a kind of "depth perception" similar to the way two slightly different sets of information combine to give us binocularity. But in this case, "binocularity" involves multiple senses, including three sets of faculties: kinesthetic, those by which we carry out physical activities; proprioceptive, related to stimuli produced and perceived within us; and temporal, which gauge and regulate our internal sense of time. Such tangibility isn't automatic: replace the 200-year-old brick wall with brick wallpaper and the rich associations between the senses immediately retreat and the experience flattens out.

The consequences of reducing experience in such ways, now ubiquitous in our digital world, have hardly been noted during the rush to turn every surface into a screen. The United States Air Force is an exception and is very aware of the effects of losing depth of experience, as revealed in a *New York Times* story:

> On its face, it seems like the less stressful assignment. Instead of being deployed to Afghanistan or Iraq, some pilots and other crew members of the U.S. military's unmanned Predator drones live at home in suburban Las Vegas and commute to a nearby Air Force base to serve for part of the day. They don't perform takeoffs and landings, which are handled overseas. But the Predator crews at Nellis Air Force Base in Nevada "are at least as fatigued as crews deployed to Iraq," if not more so, according to a series of reports by Air Force Lt. Col. Anthony P. Tvaryanas [and a] survey [that] also showed that Predator crews were suffering through "impaired domestic relationships."
>
> Why is this? Part of the problem lies in what Tvaryanas calls the "sensory isolation" of pilots in Nevada flying drones 7,500 miles away. Although there are cameras mounted on the planes, remote pilots do not receive the kind of cues from their sense of touch and place that pilots who are actually in their planes get automatically. That makes flying drones physically confusing and mentally exhausting. Perhaps this helps explain the results of another study Tvaryanas published with a colleague in May, which examined 95 Predator "mishaps and safety incidents" reported to the Air Force over an eight-year period. Fifty-seven percent of crew-member-related mishaps were, they write,

"consistent with situation awareness errors associated with perception of the environment"—meaning that it's hard to grasp your environment when you're not actually in it.[7]

Looking at a screen isn't the problem as much as the disconnection between senses. The eyes might be "flying," but not the rest of the body.

Let's for a moment revisit the hypothesis that the brain/mind operates an internal model of reality that is continuously adjusted to match up with what's "out there" in order to remain immersed. For the drone pilot, reality is experienced as sitting in a vinyl chair in a Nevada desert trailer moving a joystick; meanwhile, the pilot's internal model is trying to absorb and accept the airborne scenario as "reality," since it constitutes the arena of greatest concentration and tension. It seems that whenever there is such a high degree of unresolved discord between reality and model, there's unease; contrast that to the mind enacting itself with and through unifying sensory currents of craft, material-based work, material imagination, and enveloping environments.

MATERIALS AND REPRESENTATION

We navigate through such simulated conditions on an almost daily basis with varying effects, mostly on the unconscious. Some materials can be simulated so well that we don't notice; we even take immediate pleasure in being fooled by such special effects. Whether faux fur, Madame Tussaud's, or Las Vegas, entire industries are organized around the instant and short-lived pleasures of simulation (figure 8.4).

Material in everyday experience, though, is more visceral than imagery alone; the behavior of material affects us more than its appearance, even if its behavior is understood from that appearance. Take, for example, two kinds of brick sidewalk, one of stamped and painted asphalt, an equivalent of the instant simulation described above, and the other made of nineteenth-century brick pavers (figure 8.5).

A look at the former might not engender a second thought, but attentiveness to the latter will. The inadvertent marks of the makers, of the hands that handled the wet clay, can be seen in the image on the right, a

8.4
CNN's Anderson Cooper and his wax double.

8.5
Sidewalks made of simulated brick (left) and real brick.

memory of the lives that formed those bricks. The moss growing between the bricks forms an unanticipated symbiosis of inert and living matter. The bricks, slightly uneven, skin-like, gently accommodate the pushing of tree roots below without cracking or failing. As the years have passed, the bricks have worn ever so slightly under thousands of footsteps and the passing of the seasons. Underfoot, the fired clay gives a distinctive hard yet muffled sound and feel, even through shoes; and as residual warmth rises from the sidewalk on a cool June evening, it is a pleasant reminder of what had been a fine sunny day.

The stamped asphalt does not engage our attention in the same way. It has little capacity to hold time and life. It wears away in a few years, not like an old, weathered wooden handle but in a joyless, anonymous manner. It holds neither memory of its making nor accumulated value other than a slow-motion revealing of the material falsehood upon which it was based.

The hyperfocus on the ocular coincides with enormous demand for materials that appear new and never change their appearance over time. It is as if image has triumphed over substance, as if the material narratives of wear and aging have lost their meaning and value as representations of life's inevitable truths. The environmental incongruities suffered by the drone pilots aren't any different; they are just more extreme than in the faux-material-laden environments we experience in everyday life. The brick wallpaper, the fake tile linoleum, "wood" laminate, vinyl siding—these and all such impostor materials rupture the haptic from the visual and experience from expectations.

The authenticity or falseness of materials is not absolute; such judgments are extensions of our own moral and ethical questioning. Materials—as much as they are physical and sensory—represent things, thoughts, and states of mind. We think through materials, and in doing so make them both meaningful and reflective of ourselves: a gait is *leaden*; a gaze *steely*; differences are *papered* over; truth is a *veneer*; demeanor is *stony*; a chin *chiseled*; a grip *leathery*. A "wooden smile"—meaning stiff and without life—also projects a sense of inertness back onto wood. Representation and the senses play against each other, with representation (in the form of language) being less prone to being overturned: though a craftsman learns more of the nature of wood, its living qualities and ever-changing swelling and shrinking character, the wood of the "wooden smile" will remain as rigid and unchanging as ever. The semantics of material are slow to change; after all, language functions because it is stable and consistent.

A placard outside St. Johann Church in Davos, Switzerland (figure 8.6) states that its builders 700 years ago built the steeple true and straight.

8.6
St. Johann Church, Davos, Switzerland.

Soon after completion, though, it began to twist clockwise. Why? One suggestion, admittedly in jest, is to blame it on the Coriolis effect, the deflection caused by a force acting on objects in motion relative to a rotating frame of reference (think water down a drain and hurricanes).[8]

But were that the answer, all the twisted steeples of Europe would rotate in the same direction. They do not: There is as much clockwise rotation as there is counterclockwise. Another theory is that all these steeples were twisted "by design"—in other words, deliberately built that way. It is difficult to prove, especially since these steeples have all been rebuilt and restored. Very little, including their cladding, is original.

The culprit behind the rotation of the Davos steeple, at least according to on-site information, is the green wood structure itself as it dried and shrank. Plausible? Whether one is seeing the structure in photographs or viewing it from the ground at the church, such forensic inquiry is difficult at best. It has been said that a pastor in New Jersey, speaking of the problems with his own church's steeple, suggested another possibility when, after a tremendous windstorm, the tower had to be replaced because it had become twisted. The possibility of external wind forces contributing to the twist is compelling because it allows for clockwise or counterclockwise results while not discounting the internal force resulting from shrinking timbers.

Sunflowers are a good example of how twisting might be the result of two simple forces, one internal and the other external. Sunflower seeds grow at a certain rate according to genetic instructions (internal forces). As they grow, they bump into each other and are forced into a twisting geometry (external forces).

The steeple of the church of Notre-Dame in Grand-Marchin in Belgium was one of about forty or so twisted steeples of Europe before it was destroyed in a fire. Despite its obvious "flaw," a decision was made to match the former steeple, to transform the twist into design, when it came time to rebuild in the same timber technique (figure 8.7). It was a remarkable moment in which material behavior was transformed into architectural "language," the syntax now purely synthetic, denoting something the new steeple is not: a twist formed through time, material, and force.

8.7
New steeple of the church of Notre-Dame, Grand-Marchin, Belgium, during fabrication.

In chapter 5, I described René Descartes's famous experiment to demonstrate the mind's conceptual hold on material that allows us to observe wax melt and understand it is still the same material. The steeple at Grand-Marchin demonstrates the exact inverse: the material (twisted wood) remains consistent (by being reconstructed), but the mind has changed its conception of the twist; the new, duplicate steeple is now a symbol of history, loss, and revival. That doesn't necessarily negate the wax experiment, but it does refute the idea of the mind having exclusive domain over the conceptualization of material.

In the context of making, material is integral to bodily experience, reflective of human and cultural values, and, as in the case of the church in Grand-Marchin, it undergoes a transformation from a physical to symbolic form. What role do material and material imagination play in the broader mediation between the individual self and the world? To answer that question, let's explore material within the formal and spatial context of a very famous house.

WHAT THE TUGENDHAT HOUSE TELLS US

Every house is a form of correspondence between someone's interior self and the physical world, and hence has something to tell us about material imagination. With so many givens in most domestic environments, however, the material imagination does not typically manifest itself in ways that become exemplars of the concept. The architect Ludwig Mies van der Rohe's Tugendhat House (Villa Tugendhat) in Brno, Czech Republic (then part of Czechoslovakia), is an exception. Commissioned by Fritz and Grete Tugendhat and completed in 1930, the house was an uncompromising experiment in living. The architect, widely considered one of the pioneers of modern architecture, was given complete control of the design of the house and every piece of furniture.

Mies radically reconceptualized the concept of *house* into a metaphysical journey from the physical world into a dream state of appearances. To appreciate this, some background explanation is needed from philosophy—first from two German philosophers, Nietzsche and Arthur Schopenhauer (1788–1860), from whose work Nietzsche draws in the excerpt below. (Note that the principle of *individuation* to which Nietzsche refers is a term, used most commonly in philosophy and psychology, having to do with one's definition of one's self as separate from others.)

> In an eccentric way one might say of Apollo what Schopenhauer says, in the first part of *The World as Will and Idea*, of man caught in the veil of Maya: "Even as on an immense, raging sea, assailed by huge wave crests, a man sits in a little rowboat trusting his frail craft, so, amidst the furious torments of this world, the individual sits tranquilly, supported by the *principium individuationis* and relying on it." One might say that the unshakable confidence in that principle has received its most magnificent expression in Apollo, and that Apollo himself may be regarded as the marvelous divine image of the *principium individuationis*, whose looks and gestures radiate the full delight, wisdom and beauty of "illusion."[9]

The Enlightenment was a triumph of reason: the belief in an existence subjected to a mechanism of laws and order, through which humankind could grasp the world as a knowable system. The individual was nothing more (or less) than a rational being, fully integrated into the new universal model created by applying scientific reason to economics, social order, political systems, and so on. In reaction to this model, the German romantic movement was born of a sense that the self was forever apart from the world, longing for the recovery of a lost unity. It arose in response to the widely held view from the Enlightenment that reason and pure logic could explain almost everything, that notion being a rejection of the irrationality and ignorance of the Middle Ages (what Enlightenment thinkers called the "Dark Ages"). The romantics, in contrast, celebrated the spiritual yearning of the Middle Ages, emphasized intense emotion, and glorified nature, and in their work expressed a longing to recover unity between humans and their world. In Germany, romanticism was the dominant intellectual movement from the late eighteenth century until the middle of the nineteenth century.

The German romanticists hoped that the productive imagination could counter the powers of reason.[10] In this context, the Apollonian and Dionysian impulses can be considered romantic urges to overcome all distance, to be one with the all, either through "drunken dancing" or dreaming. Dreaming in this respect is the opposite of dancing; rather than dissolving estrangement through the action of the body, the dream is nothing less than the transformation of the physical world into pure image. The waking version is material imagination.

This transformation can occur through dematerialization or by distancing oneself from the physical world and/or removing a middle ground. Visitors to the Grand Canyon often experience this absence of middle ground, describing their experience of first approaching the rim as looking at a "pure picture" or diorama. The German painter Caspar David Friedrich's 1818 *Der Wanderer über dem Nebelmeer* (*The Wanderer above the Sea of Fog*) also makes any middle ground absent (figure 8.8). The painting evokes the isolated soul aching for the infinite, the romantic pathos of Schopenhauer's and Nietzsche's *principium individuationis*. The wanderer

8.8
Der Wanderer über dem Nebelmeer, Caspar David Friedrich, 1818.

gazes (akin to Schopenhauer's man in a rowboat) at a scene of immensity transformed into a flattened image, the clouds below having obliterated any perceptual connection between the distant landscape and the craggy outcrop on which he stands. That dark and craggy outcrop is backlit, its hard materiality intensified in contrast to the distant mountains. The wanderer is our surrogate, seeing and dreaming for us. His figure, despite being in the foreground, is nearly a silhouette, flattening and melding with the dark outcrop. He represents the Apollonian spirit, at one with "appearance" and its beauty.

The Tugendhat House, designed more than a hundred years after Friedrich's painting and decades after German romanticism had given way to other intellectual movements, may seem to hold little in common with *Der Wanderer*. After all, this iconic house is considered one of the great examples of high modernism, functional design, and the architectural rationalization of everyday life.

The early decades of the twentieth century were a time of experimentation for artists and architects searching to define the newly modern era in society. The influence of *The Black Square* (figure 8.9), a 1915 painting by the Russian Kazimir Malevich (1878–1935), can hardly be overstated. Malevich was the originator of the avant-garde suprematist movement; his painting represented the desire for a clean break from the past, an attempt to "begin again" from nothing.

Mies's Barcelona Pavilion, contemporaneous with the Tugendhat House and designed for Germany's exhibition at the 1929 International Exposition in Catalonia, is widely considered the architectural equivalent of a suprematist work, built up from nonrepresentational primal elements floating free of gravity or physical continuity (figure 8.10).

Like any built work, the Barcelona Pavilion never escapes gravity or materiality, but it nevertheless shifts our attention toward its *image*. As historian Robin Evans observed:

> By virtue of its optical properties, and its *disembodied physicality*, the [Barcelona] pavilion always draws us away from consciousness of it as a thing, and draws us towards consciousness of the way we see it. Sensation, forced in the foreground, pushes consciousness into apperception. The pavilion is a perfect vehicle for what Kant calls aesthetic judgment, where consciousness of our own perception dominates all other forms of interest and intelligence.[11]

I mention this by way of introducing the Tugendhat House as a building that, like the Barcelona pavilion, is experienced not as a

8.9
The Black Square, Kazimir Malevich, 1915.

representation of something or symbolizing something, but rather as something metaphysical—metaphysics being defined by French philosopher Henri Bergson (1859–1941) as "the science which claims to dispense with symbols."[12] Dreams are like this: although they present us an image world laden with unusual, even floating, elements, our experience is direct, not interpretive. Only upon waking do we realize the strangeness of our dreams and attempt to understand them for their symbolic content.

The house (figure 8.11) is built into a south-facing slope along a quiet residential street overlooking the city of Brno. The plans show a lower main living level facing the garden and city and an upper level of bedrooms and

8.10
Barcelona Pavilion, Mies van der Rohe, 1929.

8.11
Tugendhat House, view from street, Brno, Czechoslovakia, Mies van der Rohe, 1930.

an entryway fronting the street. Early sketches by Mies are dichotomous: the upper level is drawn with forceful thick charcoal lines, the lower level with a few wispy strokes. The significance of this will become apparent.

A visitor first encounters the house as low, compact, even dense planar volumes of glass and stucco set slightly back behind a steel railing, a wide, paved, stone apron connecting street to entry. Stepping onto this field of stones distills the elements in view: stucco wall, horizontal roof plane, vertical mass, curved glass element, single dull bronze cruciform column, and a framed view of the city beyond. This visual reduction has the effect of heightening the senses, making even the smallest of details pregnant with meaning. Nature is distilled and almost invisible, its vestiges now controlled by the architectural order, yet its presence is acute: a few planters of topiary are aligned on the rectilinear grid of the apron's terrazzo. Small tufts seem to have been planted in exacting fashion in the gaps between the stones (seen in early photographs; figure 8.12). The view of the cityscape beyond is framed, held at a distance, and spatially flattened. The entry is demarked by the cruciform column—a faint echo of compass rose, gnomon, and ordination—heightening consciousness of the sun's play on the pavement and shadow cast by the horizontal entry roof. In sum, the adumbrated physical world of the exterior entry portends.

Entry is through a heavy, thick, hinged wooden slab of a door that fills a compressed gap between the semicylindrical glass element and the body of the house. Inside, the elements of the exterior recur but are transformed and intensified. The exterior stone surface continues into the interior, each stone the same size and proportion, but the material is now a far more figured and directional travertine. With the stones placed in opposing orientation in a subtle checkerboard, the field sustains its gridded neutrality. A single, leafy plant keeps a tenuous connection to the outdoors through the visually obfuscating opal glass. A dematerialized reflective chromium cruciform column marks the center of a semicircular stair that descends to the lower level (figure 8.13). This twilight zone, a between state, is both more intense materially and more dematerialized, wakening the material imagination. Natural light is no longer physical, shadowed, directional, or optical, but a timeless shadowless glow, as if experienced from behind the

8.12
Entry, Tugendhat House, Mies van der Rohe, 1930.

closed lid of an eye. There is no outside or inside here; all that is replaced by a suspended state oscillating between perception and representation.

The semicircular stair repeats the earlier track of rotation from outside into the entry. The winding descent is a second act of disordination as it carves its way into the solid ground, arriving humbly at the darkest corner of the main level—leaving the visitor without bearings. A strange and vertiginous space opens. Both ceiling and floor are seamless and white, as if the arrow of gravity could point up as easily as down (figure 8.14). Shadows barely register on this floating field, ordered by chromed cruciform columns that have an uncanny effect of appearing as if in tension—slender ties that prevent the floor and ceiling surfaces from floating away from each other. A closer examination of the columns gives some insight. Typically, a square column will visually enlarge when viewed diagonally. The cruciform geometry of the columns, especially with their rounded ends, creates the

8.13
Entry foyer, Tugendhat House, Mies van der Rohe, 1930.

unusual effect of visually shrinking when viewed on the diagonal, further accentuating their slenderness and sense of tension. The mirrored finish adds to this by making the columns appear to waver, seeming almost cylindrical at certain angles.

8.14
Main living space, Tugendhat House, Mies van der Rohe, 1930.

Beyond, the space itself is monumental and billowing, delimited by precise elements. Ahead, a long compressed zone, doubly glazed, is filled with topiary. An onyx wall divides the space ahead (figure 8.15), and on the opposite end of the room a highly figured wooden semicircle emerges from a glowing light wall. An enormous 79-foot-long glazed wall presents a view of the landscape and city beyond. The viewer is disconnected from the scene by being raised above the ground (the main level is a full story above the garden level). The view presented is flattened and almost surreal—a pure image pressed up against the glass wall.

Schopenhauer's man is serene in his rowboat despite the surrounding storm. His Apollonian spirit turns away from the waves and wind, trusting instead in the beauty of appearance. The man's serenity is a state of mind—not a withdrawal of the powers of perception, but rather transference of the tactile and haptic into a reverie of image. It may be tempting to think of the Tugendhat House as such a vessel, protecting, accommodating, and inspiring its occupants to dream on—but that would be wrong. Further

8.15
Onyx wall, Tugendhat House, Mies van der Rohe, 1930.

examination of the particulars of the house, including its materials and details developed with such syntactic precision, leaves little doubt of a greater aim for this *Gesamtkunstwerk* (total work of art): for it to embody an Apollonian *Weltanschauung* (worldview).

The house doesn't quite fit any historical model for the "new architecture." Critics and historians have described its upper level as cumbersome, unresolved, and compromised.[13] The bedrooms in particular are regarded as ordinary in comparison to the spectacular openness of the downstairs. This presumes a homogeneous spatial conception for the house, and fails to consider that the core of Mies's architectural conception is the upper level's dense physicality transforming into the disembodied spatiality of the lower level. The upper floor is composed of hermetic capsules (the bedrooms) pushed together on the gridded stone field, in marked contrast to the diaphanous spatial composition of downstairs (recall the early sketches of Mies, the thick charcoal marks used to think through the upper floor plan and the minimal marks for the lower level) (figure 8.16).

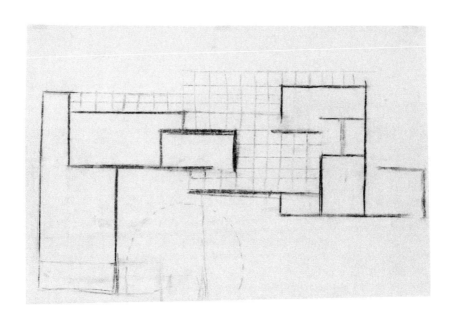

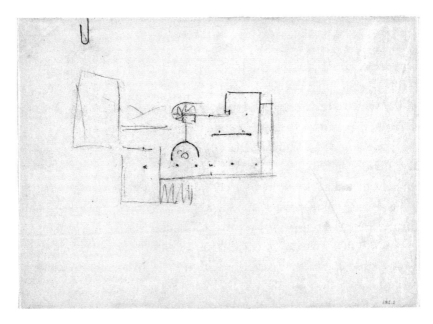

8.16
Plan sketches, upper (top) and lower (bottom) levels, Tugendhat House, Mies van der Rohe,
1930.

Here, material imagination becomes apparent. When drawing, you imagine a different condition with thick charcoal marks than with thin, wispy ones. In other words, the thick, broad strokes of the charcoal are the trigger—in this case, to conceive and organize and represent chunks of spaces such as we find on the upper floor of the Tugendhat House. Mies used a much thinner piece of charcoal or graphite to sketch the lower-level space, its character thus in keeping with the light touch of those marks. He understood that thought emerges through media and worked the two floors accordingly, bringing them into dialogue.

While each floor serves very different functions of the house, there are a number of elements that appear on both floors, codifying the dematerializing metaphysical transformation from the entry to the living space. Compare, for example, the informal sitting areas of the naturally lit upper foyer with its lower-level equivalent, adjacent to the electrically illuminated wall (figure 8.17). The upper chairs are of dark wicker, the lower ones of white leather; the former textured and natural, the latter ethereal and physically detached. Outside on the upper terrace, a leafy rectangular steel belvedere and semicircular seating echo the main sitting area and semicircular dining area below. These comparisons can be seen in the original photographs, in particular those by Studio Sandalo (figures 8.18 and 8.19). The contrasts couldn't be more clear. The photo of the upper terrace is the *only one with a living person*: in the background, the Tugendhat children stand naked in the sun near the belvedere. Downstairs, a stone sculpture of a woman's torso occupies the analogous space, the main living room sitting area in front of the onyx wall. Original Sandalo photos reveal the downstairs bereft of a single living plant. Supplanted by cut flowers, the living has been replaced by the image of life.

Quietly, these pairings define the relationship between inside and outside, not as a question of living in nature but of the nature of living. Both transition zones between outside/nature and the inner house evoke the natural and invite comparison. Mies had spoken of the transition from out to in as different than from in to out. The house treats the former as memory and the latter as imagination. The upstairs area strips away, leaving the residual substance of nature—the sun's glowing light, textured travertine,

8.17
Upper foyer (top) and lower-level (bottom) sitting areas, Tugendhat House, Mies van der Rohe, 1930.

8.18
Upper terrace, Tugendhat House, Mies van der Rohe, 1930.

the single plant, and the passing body's inscription into the carved stairway. Downstairs, the long eastern conservatory puts nature under glass. Here the outside is not lost; rather, it collapses against the topiary of the winter garden. Nature appears semi-fossilized, almost marble-like—an image (figure 8.20).

The conservatory wall mediates between a vital and an ossified nature. Directly adjacent, the iconic onyx wall, which divides the study from the living area, marks the completion of the transition of nature from the

8.19
Main living room sitting area, Tugendhat House, Mies van der Rohe, 1930.

substantive to the symbolic. The chairs of the living room, too, are significant: two green "Barcelona" chairs face three white "Tugendhat" chairs, as
if to symbolize the individuality underlying a lively conversation.

Architecture derives much of its meaning from the norms by which
it operates. A "normal" living room is furnished for conversation, and (in
temperate climates) is traditionally associated with archetypal fire, the
hearth, and dwelling. To suggest a dream state within the norms of everyday reality requires something akin to a metaphor—the displacement of
the perceived condition with an imagined one. By virtue of its placement
and appearance, the onyx wall of the Tugendhat House—one of the most
discussed stone elements in all of architecture—triggers associations with
fire, not so much the image of fire but the metaphysical significance of fire.
Standing where the fireplace would have been, the onyx wall's swirling min-
eralized patterns evokes a similar state of half dreaming or reverie resulting
from the tranquil contemplation of actual flames or mesmerizing patterns.

8.20
View toward conservatory, Tugendhat House, Mies van der Rohe, 1930.

On this topic, the Swiss-German artist Paul Klee (1879–1940) observed, "In the restaurant run by my uncle, the fattest man in Switzerland, were tables topped with polished marble slabs whose surface displayed a maze of petrified layers. In this labyrinth of lines one could pick out human grotesques and capture them with a pencil. I was fascinated with this pastime. My bent for the bizarre announced itself."[14]

Despite its material intensity and density—it is stone, after all—the onyx's polished and reflective surface in the Tugendhat House assumes a diaphanous lightness and immateriality, altering its appearance when seen from various vantage points and in different light conditions (like the Klee tables). And even more phantasmagoric is the fact that the stone transmits light: "Mies himself supervised its sawing and the assembling of the slabs in order to make the most of its grain. However, when it turned out afterwards that the stone was transparent and some parts on the back shone red as soon as the sunset illuminated its front, he, too was surprised."[15]

Such paradoxical properties make the onyx wall otherworldly. It eludes stability, yet presents a timeless character. Its materiality is both intensely present and completely absent when seen only as image (figure 8.21).

In German, two very different words are counterparts to the English word "wall": *Wand*, typically used for the wall of a room or building, with spanning and vertical construction that is gravitationally neutral, framed or clad or woven, and neither weight-bearing nor weighty; and *Mauer*, as in the Berlin Wall or the Great Wall of China, a heavy weight-bearing wall of massive construction made of stones or brick or some other similar material. The two columns immediately adjacent to the onyx wall are the *Mauer*, bearing the loads and liberating the onyx to be the weightless *Wand*, which is further freed from any framework or tectonics, floating almost miraculously in its fabric-like thinness, as if Medusa had changed the living into stone. Throughout the house, the coupling of a hardened and residual physical nature with an expansive, weightless, and dematerialized spatial field charges the whole with mystery, defying casual normalcy.

8.21
Onyx wall and columns, Tugendhat House, Mies van der Rohe, 1930.

The two conditions offer no resolution, but instead manage to catalyze a sense of quietude and mental stimulation—as attested to by the owners in rebuttals to critics questioning the "livability" of the house: "But the way it is, the large room—precisely because of its rhythm—has a very particular tranquility, which a closed room could never have. . . . We love living in this house, so much so that we find it difficult to go away and are relieved to leave cramped rooms behind us and return to our large, soothing spaces."[16]

The wanderer and the sailor in the storm dwell in immensity, which for the romantic spirit is the feeling of unity with the world. Likewise, the house has stripped away the comfortable "middle," leaving only extremes.

Vision, from a physiological perspective, is inherently rational, tending to make sense of things by fitting them into what we know. Removing the middle affects sight, inhibiting rational vision and thus opening imaginative vision—the tendency to see "into things." The tactile dimension similarly dispenses with the "normal middle" of materiality, leaving immaterial conditions such as the white linoleum flooring and the cruciform columns or incredibly rich ones that include the special woods used throughout the house. Downstairs, in particular, the semicircular dining area made of Makassar ebony appears unearthly seen against the white linoleum floor. Great care was taken to avoid patterns or symmetry in the veneers; otherwise, the wall would have had a pictorial appearance. Instead, the wall is reduced to the purely elemental and, like fire, gives rise to material imagination.

The cruciform column is the most emblematic element of the Tugend-hat House, not by appearance but in its cross-section, which amounts to a diagram of the house's spatial and metaphysical order (figure 8.22). We find a hardened physicality in the four interior iron Ls riveted together to form the cruciform, which shrink in contrast to the chrome covers. These blossom outward from the ironwork, finding form in their geometrically idealized profile, distinct from the tough squarish iron Ls, and thereby reiterating the *Mauer/Wand* distinction and the transformation of physical nature into pure appearance.

Several details of the long glass wall, arguably the house's focal point, make clear the degree to which both architect and client were aware that

8.22
Cruciform column plan, Tugendhat House, Mies van der Rohe, 1930.

its transparency was not intended to make a physical connection to nature, but rather to create a kind of splendid isolation. As Grete Tugendhat wrote, "The connection between interior and exterior space is indeed important but the large interior space is completely closed and reposing in itself, with the glazed wall working as a perfect limitation. Otherwise I, too, would find that one would have a feeling of unrest and insecurity."[17]

A blurry photograph of the great window expanse, with a silhou-etted contemplative Mies (figure 8.23), reprises the wanderer from Caspar David Friedrich's painting. The white linoleum floor, like the sea of fog, contrasts with and accentuates the disconnection from the scene of the city of Brno beyond. The raised platform of the house, whose clifflike edge drops off at the window expanse, effectively removes the middle ground, transforming the view of Brno into an image. A small detail, or rather its absence, heightens the Apollonian sense of being-in-the-world. The railing along the glass, which appears in photos taken six months later, is missing, leaving us to wonder: Did Mies intend for the absolute condition of the railingless opening? After all, he had argued with the Tugendhats themselves against window awnings, which as a compromise were designed to vanish completely behind the exterior fascia.

Despite the contrast between the interior and exterior, the scene is one of unification, a metaphysical construction made of the displaced and dreamlike floating pieces of rich physicality fused with the image of nature. The transparent plane isn't an agent of continuity; rather, it acts more as a membrane against which the exterior has projected itself, and which the imaginative vision of the interior, in turn, has appropriated for its own contemplation. This apotheosis of the Apollonian has a profound influence on the inner self: expanding it, pulling it out from its usual corporeal

8.23
Ludwig Mies van der Rohe visiting the Tugendhat House, 1931.

"home" to dwell disembodied in the interior, comparable to dreaming, distinguished by its "out-of-body "character.

In chapter 2, I wrote about the earliest drawings that are found in almost inaccessible recesses of caves. There is little evidence that these caves served any other function at the time of these drawings, a finding that supports the "sacred practice" argument.[18] To draw a line between these painted caves and the interior of the Tugendhat House may be dubious, but if we see both as models of the world, models of reconciliation with our own inner disembodied spirit, parallels begin to emerge. The images of animals and the hunt on a cave wall and the scene of Brno pressed on the glass at Tugendhat collapse distance—between human and animal worlds in the former, and between self and modernity in the latter. In both, images serve as interpretive screens on which their liberated signs and symbols are available to make sense of reality. The flatness of images loosens their content—not only can they be experienced, but they can also be "read" as we read a text. Such images have the power to hover simultaneously as appearance and idea. They can enter our mind as words do, and our mind can enter them as only dreams do.

9 MATERIAL AND *UMWELT*

"Therefore," wrote Walter Benjamin (1892–1940), the German Jewish philosopher and cultural critic, "I am distorted by resemblances to everything that surrounds me here."[1]

Why do we seek a certain place in order to accomplish something—to read or write or sketch or empty our mind? Such an understanding of how our surroundings inform our mental state comes naturally and is uncontroversial, but as this book explores, quantifying this relationship remains elusive. I may be certain that my material surrounding, or *Umwelt*, participates in my thoughts, but I cannot say why or how sitting here will affect my mind.

Umwelt, the German word for "environment" or "surroundings," has come to mean our "self-centered world,"[2] that is, the sum of what we each can potentially sense in our surroundings. This concept was first introduced in 1909 by the German biologist Jakob von Uexküll (1864–1944) to express the observation that different animals in the same ecosystem respond to different environmental signals.[3] His work caught the attention of semioticians (semiotics is the study of meaning-making through signs, analogy, metaphor, likeness, etc.), and the *Umwelt* concept was expanded, most notably by the Hungarian-American semiotician Thomas A. Sebeok (1920–2001) and philosopher Martin Heidegger (introduced in chapter 7); indeed, it plays a central role in Heidegger's major work *Being and Time*, which is an account of human existence.[4] It is a concept of a surrounding environment bounded by what we are aware of or what affects us directly.

Material in our *Umwelt* can range from what we can touch and smell to the sounds we hear and what we see. *Umwelt* is the material/spatial environment into which we can expand our cognition and interactions. As first proposed by Clark and Chalmers in their essay on the extended mind and as later argued by Varela, Rosch, and Thompson in their book *Embodied Mind, Umwelt* is what we self-enact with and from.[5] *Umwelt* includes the continuum of the underlying world in which we all participate without being fully aware of its cosmic nature. Its order, while not necessarily in harmony with our own conceptions, unites what is known in common understanding while remaining beyond consciousness.[6]

Some areas of culture, such as poetry, literature, music, dance, and the fine arts, are attuned to this cosmic order, while other fields are more self-referential, constructing orders that harmonize to a greater or lesser extent with the *Umwelt*. Design, economics, medicine, computer science, and others are valuable—even essential—constructs, but their occasionally disastrous outcomes reveal how far removed they are from the larger and deeper unity.

I have discussed how the mind is influenced when we directly work media and materials. *Umwelt* includes such engagement, but extends beyond to the latency of the surroundings. Can we discern how our surroundings interact with the mind? After all, the power to shut things out has made us confident in our independence from the influence of all manner of distractions, including those of our *Umwelt*.

The perception of the floor creaking down the hallway cannot be reduced to acoustical sensations. We cannot separate the sound from its totality. The creak is something our receptive mind moves toward and meets halfway (you hear it "over there," not in your head). A complex and concrete phenomenon forms, part perceived, part constructed in mind, and part bodily motor response—a building up of spatial, acoustic sensations, a whiff of a smell perhaps, almost imagined, named, recalled, perhaps sentimentally, tacitly remembered, the quality of light and image and sound mingling into a vast and rich brew enmeshing the mind with this part of the *Umwelt*, a form of embodiment that arises from the limiting and resisting material surroundings. Put another way, the presence of the wholeness of experience cannot be isolated from its spatial/material context and is, in fact, embodied in it. That is why what we say is "I hear the floor creak," not "I am receiving interesting acoustic sensations." And it is why we shouldn't assume the mind operates without influence from what surrounds it.

Philosopher Henri Bergson probed how the mind and experience come together. In his view, the latter is modulated by the degree of one's familiarity with it: a brand-new experience is raw and mostly sensed, while the familiar combines the many memories of that experience with the perception of what he calls "memory-image": "any memory-image that is capable of interpreting our actual perception inserts itself so thoroughly into it that we are no longer able to discern what is perception and what is memory."[7]

Bergson believed an experience, such as the creaking we hear, is our memory (or, in the case of visual recollection, memory-image) joining with the sensory perception of the creaking. When we hear a creak for the first time, it is sensed; when we hear it for the umpteenth time, it is a memory-sense.

To link physical action with perception, Bergson also used the example of walking through an unknown city for the first time. At first, there is doubt at each street corner about how to proceed, bringing awkward hesitancy to movement. But after some time, as new memory and perception fuse with bodily action, familiarity sets in and ambulation through the city becomes smooth and mechanical. A spatial matrix of the city forms. But where is this matrix located? Is it in the mind as a memory or a mental map?

Allow your memory to take you to a familiar city and recall how you have navigated through it. There are fragmentary sensations of direction, half-formed images of key landmarks, a plan perhaps, but without any clear path or geometry, just the sense of shape, and a few organizing lines—such as a major avenue, a neighborhood, a river. When you return to that city, familiarity rushes back. You recognize every street corner, and movement again becomes fluid. Our spatial matrices reconstitute themselves from mind (memory) *and* the situation, the *Umwelt*, which participates in our mental state so completely that we are unaware of it as "other."

As an organizational strategy, this makes a lot of sense. Why would we mentally burden ourselves with a rigid, overly articulated mental map that would be thrown off by the slightest incongruity? Instead, our "half matrix" of the mind (the other half being the activating agency of reality) is an amalgam of blurs and smudges, less specific but far more plastic and adjustable to changing circumstances. This points to an inescapable conclusion that the mind is found as much *outside* the brain case as within it. A corollary to this is that our physical surroundings both give agency to and are the agency of the mind, which enacts itself by coupling with our *Umwelt*.

When our surroundings become familiar, we "inhabit" them, we become habituated, we become—paradoxically—both immersed and liberated. As perception becomes subconscious, our mind tends more to its internal dialogues and thoughts, seemingly oblivious to the surrounding *Umwelt*. Who hasn't experienced a "wandering" mind, losing hold and attention to the here and now? Between these two situations, the unfamiliar and the familiar, lies a certain transitory state where perception is still alert and conscious but movement and experience become immersed and in tune with the *Umwelt*. In this state, the mind balances between independent consciousness and being strongly receptive and attentive to the *Umwelt*. As described by Bergson, something is still incomplete, mismatched, or ajar between memory-image and perception. Imagination resides within this special state; the ability of the mind to be triggered by and to move amidst the disjunctions between memory and perception makes for "leaps of the imagination."

Material's influence on the imagination is not limited to that which is physically proximate; as part of the larger *Umwelt*, materials at all ranges of distance are perceived, even haptically (I can sense the coolness of the stone surface without touching it) because perception is itself directly coupled to memory, including the memory of touch. And Bergson noted that perceptions are tied to actions as memories are to dreams. Imagination seeks to mingle actions and dreams; when "what has been" doesn't match up with "what is," it triggers a movement toward "what could be."

Our perceptions and interactions with material change with its situation. Comprehension and the conceptualization of material are contingent upon proximity and the totality of the *Umwelt*. The potential for our memory/perception to mismatch with material can be placed in roughly three zones of influence. Mismatches can arise from the direct manipulation of material; between multiple modalities of material engagement; and, when material is at an out-of-reach distance, from the interactions between the senses and movement.

Let's begin with material within reach. In previous chapters, material resistance has been central to the discussion of creativity and the imagination. Manipulating material is the most direct experience of the physical environment; there, the mismatch between perception and memory can be understood as the innate capacity of material to defy familiarity and expectations consistently. Material manipulation is a provocative activity because it prolongs the valuable transitory state between familiarity and unfamiliarity.

Just because material is beyond reach, however, it isn't beyond affect. The creative processes of the arts and design involve a constant interaction between indirect forms of engagement, namely media/medium, drawing, and language. Such a toggling between modes also extends the transitory state, generating the slight mismatches that inspire imagination. Words don't line up with a model or object and a drawing doesn't align with either, giving impetus to make imaginative adjustments in every modality of work.

At distance, material engagement is limited. A far-off mountain may be visually perceived, but its material characteristics seem too remote to be of any significance. The stability of such a distant backdrop would seem to limit the possibility of mismatches between perception and memory that animate imagination. But here, too, a transitory state between unfamiliarity and familiarity can be found. Everyone feels the profound sense of immersion in the immensity of an approaching mountain range—a fullness of experience not replicated in any image, despite the seeming pictorial flatness that distance engenders. The materials of the atmosphere that animate the day—clouds, haze, and other materialities that comprise weather—and the shifting seasons provide structure for the mind's extension and grasping of space, providing some of the mismatch between memory and perception of great distances.

Ultimately it is movement, bodily movement, that draws the far-off mountains into the personal experience and feeling that, as discussed earlier, precedes and triggers thought. Movement generates subjective and immersive awareness in at least two ways, physiologically through proprioception and visually through parallax. Proprioception, the sensation of one's spatial self, relies on bodily sense (position and muscular feedback), vestibular sense (balance), and visual perception. With parallax (defined in chapter 6), the horizon bobs up and down. Backgrounds slide about as we focus on a foreground. The further space extends, the more parallax plays a significant role in how we experience that distance.

This explanation is an attempt to begin to answer the question of how material at a distance might interact with the mind. While the physicality of material at a distance is no longer a direct factor, our relationship with that material is not passive; we still interact with our *Umwelt* because of our moving bodies and senses in space. Sonic qualities are highly attuned with material characteristics; the absorptive wash of pine trees in a forest, or the hardness of a cliff face, are *heard* from a distance. The residual warmth radiating out from the same cliff after the sun has descended lets us *feel* the stone's thermal nature without coming close. Such a haptic sense, a mental "touching" of our *Umwelt*, arises from all the senses being overlaid with memory, not perfectly but interactively, in a back-and-forth correspondence. The movements of our body yield an extended transitory state, which speaks to humans' general love of the outdoors, including hikes and walks through natural settings.

Human experience as it relates to the *Umwelt* is a concrete experience. The world as we find it, whether as an urban streetscape or as a river flowing through a forest, is always experienced as a totality. Attentiveness lets us concentrate and push nonessential perception aside to be handled by unconscious sensory processes, but that doesn't change the concreteness of the world. Concreteness is manifested as resistance, which is tied to embodiment. The underlying resistance of the felt totality is a direct extension of the resistance material offers: resistance to manipulation, to forces, to any propositions or actions that contradict the concreteness of and inherent processes that underlie reality. To live within and with awareness of resistance is to live in an embodied fashion. Correctly calculating the effort to swim across a river requires the calculation to arise from an embodied sense of the situation. Such a calculation is impossible to make from looking at a picture.

Human intelligence stems from the inhibitory system of the brain, without which responses to stimuli would be purely reactive. The inhibitory system gives rise to deliberation, planning, and reflection—in short, to *thinking*. The enacted mind finds a similar inhibitory function in the resistance of materials and the *Umwelt*, bringing forth creative, imaginative, intelligent, and embodied engagement.

THE SITUATIONAL NATURE OF MATERIAL

The *Umwelt* is never stable. Our physical movements, narratives, and conceptualizations are constantly rearranging the relations among things, and in doing so alter our focus and attention. Italian architect and Pritzker Prize winner Aldo Rossi (1931–1997) explained this in his book *A Scientific Autobiography*: "The emergence of relations among things, more than the things themselves, always gives rise to new meanings."[8] Yet amidst this fluctuation, our perception of the *Umwelt* remains remarkably consistent. Our memories and stored images, memory-images overlaid onto perceptions, stabilize the *Umwelt* into an unquestioned concrete reality.

When something has broken out of the felt continuum, we become more aware of the web of changing relations that situate everything. A single orange on a table is "an orange." An orange alongside an apple and a banana is "fruit." The situation has changed the relationship, which goes unnoted. Now put the orange next to a rectangular block of wood that matches the orange's proportions and size. What is it? Some casting about for an explanation yields nothing, but the size similarity might begin to elicit a narrative: "Someone could carve that wood into an orange." The narrative, in effect, normalizes the *Umwelt*, allowing attention to move on from the slight dissonance between the two objects.

The situated context of things, including materials, is not limited to the relations among objects, but includes relations in time, space, purpose, meaning, and authorship. The fashion industry is driven by temporal context: a current thing becomes "out of fashion" even though the only change is the passing of time. An older car is noticed and admired because it is an anachronism. Move something from one place/space to another and it, too, is transformed: a syringe in a hospital is one thing; the same syringe lying by a curb is another. The simple power of dislocation propels the impulse to collect things: even a common stone collected from a distant shore, brought back, transforms into treasure on a shelf.

Purpose and meaning are deeply related when considering things and their situatedness. In chapter 2, I discussed chairs and wristwatches in this regard. The lack of functional pressure on their purpose allows design attention to shift to their symbolic value or meaning. This shifting between purpose and meaning can happen within an object after it is made, such as when a bicycle wheel is repurposed into a wall clock. The direction of the shift is almost always toward being less purposeful, which may be why a sense of pathos, even thanatos (the death instinct), accompanies so much of repurposing.

Turn a stool upside down and it's still a stool. Turn a urinal 90 degrees, have it rejected from an exhibition, and—by virtue of its authorship by a known artist and the particular moment in time—it becomes an iconic work of art. This happened in 1917, when the French painter and sculptor Marcel Duchamp (1887–1968) submitted a porcelain urinal titled *Fountain* to an exhibition of the Society of Independent Artists in New York City (figure 9.1). The work was signed "R. Mutt 1917." Duchamp was already a well-known and controversial figure in contemporary art at the time. Though the exhibition rules indicated that if an artist had paid the entry fee his or her work would be accepted, the committee rejected *Fountain*. Instead it was displayed and photographed at the studio of the famous photographer Alfred Stieglitz.

Today, Duchamp's *Fountain* is considered by many to be a landmark in twentieth-century *avant-garde* art. Had a plumber set the same urinal upside down in a gallery, would it have meant anything more than a plumbing fixture waiting to be installed? In the art world, authorship

9.1
Fountain, Marcel Duchamp, 1917.

confers legitimacy and meaning to things which otherwise may have little of either.

Consider the iconic *Painted Bronze (Ale Cans)* from 1960, by artist Jasper Johns (figure 9.2). The work is a rematerialization in bronze of two beer cans, one "opened" and "empty" and the other closed (only picking it up reveals it is solid bronze). Bronze is so culturally ingrained that anything cast from it automatically bestows monumentality and prestige; in this case, material (bronze), authorship (Johns), and a shift from the purposeful to the meaningful have utterly resituated a quotidian item. Had the piece been cast out of plastic, its impermanence and cultural debasement (as compared to the bronze) would have altered its significance. If Johns hadn't hand-painted the label, if he hadn't cast the pair as freestanding pieces, if he

9.2

Painted Bronze (Ale Cans), Jasper Johns, 1960.

hadn't already been an established art gallery artist, and if art critics hadn't written admiringly, we might never have come to know this work.

Remarkably, we have an underacknowledged form of thinking—I call it *situated thinking*—with which we can make sense of the situational interdependence of all things in the world. Situated thinking is constantly emerging from the interactions and changing relations between us and the materials and things around us. In contrast, Cartesian-based, objectifying knowledge systems mostly ignore the intelligence of situated thinking, despite it having been acknowledged in philosophy around the subject of interpretation.

German philosopher Hans-Georg Gadamer (1900–2002) saw the interpretation of texts (hermeneutics) as a kind of situated thinking.[9] In his view, there is no objective, embedded "truth" to be found in a text. Interpretation is based on a dialogical approach in which meaning and understanding emerge as an event outside the text, a result of discussion, disagreement, preconceptions, and from the text itself—in short, from the situation. Change the situation, and a new meaning emerges from a new and unique event.

Though Gadamer's work was focused on text, it applies to the situated thinking that occurs in our interaction with things: meaning is carried by and dwells in the event of interaction, and not within things themselves.

EXTENDED AND ENACTED MINDS

At the beginning of this chapter I introduced the concept of *Umwelt*, and I have discussed generally how situated thinking arises from the changing relations between things. This kind of thinking is more provisional and *Umwelt*-dependent than it is purposeful or methodical—the kind of thinking airplane pilots are engaged in as they land. Situated thinking comports with the idea of the *enactive* mind, one that forms itself from self and world (self-world) sensitive to and informed by the *Umwelt*. In contrast, the *extended* mind theory proposes that one's surroundings act more as a further cognitive scaffolding, functionally extending the self as needed.

I can think of circumstances in which one of these models, the enactive or extended, rings truer than the other. Most examples given in this book are situated and enactive—that is, thought and realization in the examples are triggered by or informed by the actions being undertaken or the medium being worked.

The airplane cockpit is an interesting *Umwelt* because it is specifically designed as a purposeful cognition extension device and not as a space of enactment, probably to the great relief of passengers. Cognitive anthropologist Edwin Hutchins analyzed what he calls the sociotechnical system of the cockpit, in which the pilots are but one of the functioning parts.[10] Focusing on the critical landing sequence of a 1990s-era McDonell Douglas MD-80, a commercial airplane, and specifically the coordination of landing speeds and flap settings, Hutchins observed how cognitive loading is redistributed in order to free the pilots to focus on other critical needs in the final minutes before touchdown. Before the landing sequence commences, a pilot takes a card from a deck that corresponds to the current weight of the plane, which provides all approach flap settings and landing speeds. The crew then works from the information on that card and, through callouts, reading, writing, listening, setting small "speed bugs" (little movable plastic indicators on the airspeed indicator face), and arranging the tasks over time, lowers and distributes the cognitive load.

> The memory observed in the cockpit is a continual interaction with a world of meaningful structure. The pilots continually are reading and writing, reconstituting and reconstructing the meaning and the organization of both the internal and the external representations of the speeds. It is not just the retrieval of something from an internal storehouse, and not just a recognition or a match of an external form to an internally stored template. It is, rather, a combination of recognition, recall, pattern matching, cross modality consistency checking, construction, and reconstruction that is conducted in interaction with a rich set of representational structures, many of which permit, but do not demand, the reconstruction of some internal representation that we would normally call the "memory" for the speed.[11]

Hutchins is describing distributed cognition, which in his view makes the *Umwelt of the cockpit* and its actors and their actions a single continuous networked system. The choreography of physical acts, mnemonic prompts, cross checks, and reads permits the memory for the speed/flap relations to be distributed systemwide. It is interesting to note that violating any of the strict procedures is itself not catastrophic but would disturb the system, which permits only one menu of correct actions to unfold over time; this is hardly an environment for provisional or situated thinking.

A small detail Hutchins pointed out, however, exemplifies how situated thinking is always at hand, even if held in abeyance by the procedures being followed. A special, wider "salmon bug" (so named because of its color) is moved on the airspeed indicator face to near the minimum speed required for the plane to fly. This allows the system to remember and the crew to maintain very careful control of the airspeed near touchdown, within 5 knots above or below that indicated by the salmon bug. Hutchins found that the crew provisionally used the width of the bug as a handy aid—since it covered just about the area on the airspeed instrument, covering the ±5 knots the crew needed to manage.

The engineers designing the cabin were thinking only of graphical clarity in making the base of the bug narrow enough that it would never obscure more than one tick mark of the airspeed indicator, according to Hutchins. The pilots, however, were using situated thinking; their adaptability, though not following the rules, lent resilience to the landing process, especially if something unexpected were to happen.

While writing this book, I spoke about situated thinking and the cockpit with a pilot friend of mine who flies large, modern airliners. With computers handling more flight procedures than in the past, he told me, the physicality of the cockpit has changed since Hutchins wrote about the cockpit in 1995. Rather than physical cards being selected from a deck, computers now establish the information, and instead of plastic "speed bugs" there are virtual icons on digital screens. Despite the automation of the cockpit, however, my pilot friend still observed that much piloting is physical, that it is a matter of kinesthesis, proprioception, bodily feeling, and looking outside as much as it is one of cognitive engagement with the instruments and instructions of the cockpit. He described the essence of flying as one part cognitive extension and one part enactment of the mind through physical engagement. The increased level of automation has made flying much less prone to errors of judgment but, by reducing the physical interaction of crews with instruments, has also lessened the opportunities for situated thinking, making for a less resilient distributed cognitive system, one vulnerable to forgetting the importance of the bodily sense in piloting—exactly the sensibility needed when things go awry.[12]

ARTICULATION AND RESISTANCE

I have discussed the nature of our interactions with the *Umwelt* and how sensory perception, manual action, movement, memory, receptivity, knowledge, conceptual structures, and preconceptions—in short, all the capabilities of the individual self—come to define things and their significance. Our interactions hold aloft our collective and highly situational consciousness of what we call the world. Such a drawing forth or articulation of the world can be viewed in opposition to but dependent on resistance, especially material resistance, which in its essential unfathomability grounds us while participating silently in this articulation.

To illustrate, imagine the process of quarrying a stone, moving it to a site, then dressing (finely carving) it to become part of an assembly that in turn becomes a building. Think of this as the process, analogous to language, through which we define, represent, and enact ourselves in a world of our making. At the same time, the physical difficulties of quarrying the massive rock face, the silent ground on which the stones bear, the sheer weight and hardness of each stone, are the material's resistance, which is brought forward by the act of articulation. The resistance of the material is more present because of the articulation of it.

This opposition, wrote Martin Heidegger in his essay "The Origin of the Work of Art," is a form of strife between the setting forth of "World" (articulation) and the setting back on "Earth" (resistance). Heidegger discussed the oppositional natures of World and Earth through a reference to a Greek temple: "The temple-work, standing there, opens up a world and at the same time sets this world back again on earth, which itself only thus emerges as native ground."

More specifically, Heidegger considered the formed material of the temple.

> [T]he temple-work, in setting up a world, does not cause the material to disappear, but rather it causes it to come forth for the very first time and to come into the open region of the work's world. The rock comes to bear and rest and so first becomes rock; metals come to glitter and shimmer, colors to glow, tones to sing, the word to say. All this comes forth as the work sets itself back into the massiveness and heaviness of stone, into the firmness and pliancy of wood, into the hardness and luster of metal, into the brightening and darkening of color, into the clang of tone, and into the naming power of the word.[13]

The passage speaks to the interpenetrating nature of World and Earth, articulation and resistance. By our actions on and with material, we draw out its character, and as we do, material quietly affirms its resistance—concealed, mysterious, and infinitely beyond our grasp. What is often misdescribed as "material failure" is the strife between World and Earth uncovering a truth that is greater than our capacity to understand.

The setting up of the World brings forth the Earth. That is not far removed from the formal imagination and material imagination of Gaston Bachelard: the articulation of "World" is toward consciousness, expressed in image and surface, while the resistance of "Earth" is toward the essential: subliminal, gravitational, and limbic.

Heidegger goes on to argue that building the temple changed how things came to be viewed, and how humankind came to view itself. Winston Churchill famously said much the same at a meeting in the British Parliament on October 28, 1943. The House of Commons had been destroyed in a German bombing raid two years earlier, and a discussion of rebuilding that chamber was under way. "We shape our buildings," Churchill declared, "and afterwards our buildings shape us." He was arguing for restoring the building to its old floor plan of adversaries sitting opposite each other that he saw as having been instrumental in "shaping" Britain's (then) two-party system and thus as instrumental to democracy.

The temple Heidegger describes is situational: the act of giving form, of building, of smoothing and cutting and shaping material, articulated and brought into mind meaning and significance and, in so doing, affected the meaning and significance of everything else in this articulated world.

The coupling of articulation with resistance, World with Earth, is reciprocal. The more there is an effort to articulate World, the more the Earth's resistance comes into being and the deeper we are set into it. This thesis holds as long as there is a community that actively participates in the work of articulation. But once the temple is no longer meaningful or resistant to being fully understood (when it loses its mystery), it loses its power to situate mankind, soon becoming passive and then a historic relic.[14]

The recent rethinking of the origins of agriculture support such a conception of the situational role of building. The familiar explanation has the development of agriculture preceding and leading to settlement, and is based on an assumption of a purpose-driven life, survival needs begetting survival responses. Lately, archaeological evidence has emerged of the very first signs of agricultural practice being found adjacent to the oldest known building sites.[15] These buildings weren't granaries or houses or anything that might be associated with supporting agriculture; rather, they were temples, which opens a new possibility for why agriculture came into being: the enormous physical activity required to build a temple may have forced the builders to modify their usual foraging and hunting practices, especially if the temple building concentrated a large and no longer nomadic population, putting great pressure on the available food resources in the surrounding area. What does this tell us? The act of articulation, of building the temple, changed the outlook on the world, bringing into view the concept of place, dwelling in place, and cultivation of the land. It also tells us that material (stone, in the case of the temple) can turn us from our purpose in ways that cannot be foreseen. The temple builders could not have known that their decision to build in place would lead to consequences that revolutionized life on earth.

The discovery of metals echoes the temple/agriculture origin story. Copper and bronze were first used as jewelry and body adornment long before being incorporated into tools and weapons. Why so? The search for meaning through the articulation of one's relation with the world and others, then as now, is a primary drive of humankind, one occluded and subordinated by reason and purpose. Yet, as these two examples demonstrate, the search for meaning may well be what drives purpose.

I referred earlier to meaning as defined by a sense of connection and relevance that is greater than self, which is closely related to the Greek word ἀλήθεια (aletheia, the literal meaning of which is "the state of not being hidden" that came to be translated as "truth," "unconcealedness," "disclosure," or "bringing into presence"). The work of articulation, and the resulting resistance, are what bring meaning into being; aletheia makes clear that meaning is a result, not something created independent of its representation. We can thus understand the work of the temple builders as an articulation of materials and space to unconceal or bring into presence that which could not yet be known. Agricultural practices would also be an articulation of world, an unconcealment of order and presence not otherwise known. The effort to use metal as jewelry makes more sense as an act of meaningful unconcealment of the world than as a form of representation such as adornment or fashion.

CULTURED STONE AND SOCIAL STONE

The work of unconcealment unfolds in technology, the arts, and cultural and political practice. What role does material have in unconcealment? I have discussed the articulation of materials into form that reveals and brings to presence, but not with any specificity. Is it possible to compare two works of differing materials to see what role material has in unconcealment?

Consider two stone-building cultures, the Inka (Inca) civilization with its famous masonry work and the peasants of the nineteenth-century city of Matera in southern Italy, whose masonry work is much less known. They are products of complex, vastly differing sociopolitical and economic forces playing out in different centuries and through very different material sensibilities. If we compare their respective masonry constructions, can we find in them a deeper unconcealment guided by the traits of the materials themselves?

Little is known about how exactly the Inkas and their predecessors built their fabulous polygonal stone walls. These walls were but one of at least four types that included more regularly shaped masonry laid in horizontal rows. The polygonal walls were reserved for the most important constructions, such as seats of power, and therefore are understood to be of the highest cultural value (figure 9.3). The fact that the Inkas commonly used the regular masonry elsewhere reinforces the understanding that polygonal masonry was not only of great prestige, but also probably required a far greater investment of skill and labor. How, then, did the material—the stone—participate in and influence thinking?

The conventional historical framing is that the Inkas manipulated material to conform to or represent their ideas and orders, but not necessarily to have those ideas be activated by the material being worked. The question cannot be answered unequivocally, but any reciprocity between the resistance of the material and the desire to manipulate it would be an indication of stone's influence.

9.3
Polygonal Inka stonework wall.

The uniqueness of each polygonal stone in Inka walls aligns with the Inka belief that individual stones, each with an essential but latent animate capacity, had the potential to transform from or into humans. In general, the hardness of the stone made shaping it challenging; the shaping of each stone to fit its uniquely shaped neighboring stones with great precision made the task exponentially more difficult. If building a wall was purpose-driven, in terms of effort this was an inefficient means to an end.

Historian Carolyn Dean has proposed that the dressing and working of stones was part of the deep Inka correspondence with their rocky and mountainous *Umwelt*, one that constantly negotiated the relationship between the natural and the ordered (articulated) world and always through the medium of stone.[16] Inka stone walls typically incorporated natural features and even grafted themselves to outcrops, implying a more situational relationship, a real back-and-forth dialogue between order and nature, the Inkas ultimately seeking to fuse the two. Their ability to build willfully, to control stone's form, was in sharp contrast to their focus on making each stone receptive and acquiescent to its context—whether it be an adjoining dressed stone or a natural rock feature (figure 9.4). This is evidence of a culture guided by the impulse to unconcealment by honoring the reciprocity between articulation and resistance, receiving and thinking from material conditions while acting on them.

The polygonal stonework exemplifies this reciprocity. The perfect fit of the stonework is paradoxical: it demonstrates total willful control on the

9.4
Polygonal Inka stonework integrated into natural feature.

part of the masons; yet in taking on form, each stone completely submits to its neighboring stones. In working this way, the Inkas deepened their correspondence with their *Umwelt*: they did not "tame" nature so much as become conjoined with it. The difficulty of the polygonal construction would have been a worthwhile investment insofar as it constituted a process of fusing humankind with nature.

Dean observed that each stone's finish, a visible result of the slow work called "nibbling" in which each stone is hammered to remove small chips, made each stone an act of becoming civilized, understood more as continuous process than product (figure 9.5).

Further to this notion of reciprocity between the natural *Umwelt* and ordering impulses of the Inkas was the practice of hauling enormous stones great distances, with as many as 2,500 men sometimes required to haul a single giant stone. There were easier ways to use nearby material, but this was a choice, perhaps because such a process of engagement with the landscape gave order to Inka society. Marshaling so many men, an achievement

9.5
Inka stone, closer view.

in itself, would have necessitated a high degree of political and social organization as well as vastly increased agricultural production, which in turn may have fostered the need to build more stone retaining walls for farming terraces. Such a self-reinforcing cycle of activity was directed as much *toward* stone as from or by stone, constituting a simultaneous seeking to order and submission to the *Umwelt*.

Underscoring this notion are cases of massive stones being hauled almost the entire way to their destinations before being abandoned as "tired stones." Religiosity aside, this act goes against the willfulness of hauling a stone, making abandonment of a stone an act of attentiveness and submission to the greater landscape—which the Inkas believed held animus—and to its material elements.

Contrast the cultured stone of the Inkas with Matera's social stone. Matera and its surrounding landscape in Basilicata, Italy are one of the oldest continually inhabited places in the world. The region's natural caves, part of a geological process associated with the porosity of the local sedimentary rock, have long provided ready-made shelter for troglodytes, passing travelers, and pilgrims. Over time, these caves were modified by excavation and building to accommodate more formal uses such as rupestrian churches (that is, those composed of or on rocks, as in caves). By the nineteenth century, the complex ravined terrain of Matera itself became subject to intensive cave excavation and additive building, resulting in a rather astonishing morphology, fusing landform with building, entangling subtractive and additive construction processes (figure 9.6). This could happen only because of the malleability of the landform material, a porous limestone so soft that it could be cut with a simple handsaw or dug with the most basic of tools.

The softness of Matera's stone offers little resistance compared to the granite of the Inkas, making the local stonework vulnerable to weathering and allowing it to be easily altered and repaired. The medium of soft stone has influenced the sensibility of buildings, giving them a slightly provisional air without ever aspiring to heroic types of construction. Given that any investment of energy can literally be rubbed away, there is little reason to freeze or monumentalize in such a fungible *Umwelt*.

9.6
View of Sassi di Matera, Basilicata, Italy.

Though such an acquiescent medium offers less resistance to the hand, it doesn't deny unconcealment. In the case of Matera, articulation and resistance are displaced away from the stone itself, but nevertheless transmitted or unconcealed through it. Were you to visit, you would immediately notice the very dense settlement of the ravine/cave areas (locally called *sassi*) and the extreme articulation of social orders and the spatial pressures that come from proximity. The excavating and constructing activity in the area continued apace during the nineteenth century until physical limits were reached. Caves could be excavated only to a certain width or would risk collapse or transgression into a neighbor's cave. With no command and control, planning, or formal codes to assert authority, a cellular, organically organized world emerged, articulating a complex social order within the substrate of the amorphous subtractive and additive landscape (figure 9.7). The resistance of Earth appeared when limits were reached, proximity became extreme, and the soft stones deteriorated with the passing of time.

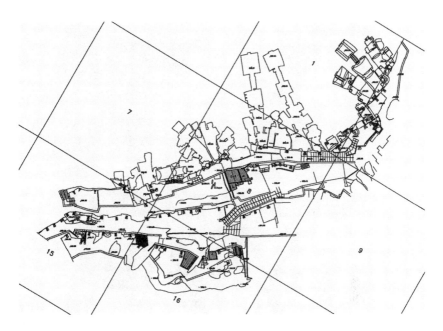

9.7
Plan of Sassi.

Similarly to how clothing is a fine-grained articulation of social norms and orders, the soft stone of Matera was suited for spatially articulating the relations between family, neighbor, realms of privacy, and shared inhabitation. An example of such a fine level of articulation can be seen in a small courtyard that constituted a neighborhood (*vicinato*) of more than a hundred inhabitants living closely together with few means and no running water (figure 9.8). Two staircases share a common dividing wall/rail between two properties/families. At the base, the wall is low, permitting the stair to merge with the courtyard. Ascending, the wall rises relative to the stair until it separates one stair from view of the other, finely calibrating the degree of privacy and subtly modulating the familial and neighborly relations of the *vicinato*.

The wall is not a representation but rather an articulation of social order to which a soft, "easy" material can respond. If the stone had been much harder, requiring far greater skill, time, and effort to work, would

9.8
Courtyard, Matera, Basilicata, Italy, 1951 (photo by Henri Cartier-Bresson).

something as situationally transient as neighborly relations been given consideration?

André Kertész's famous 1977 photograph of MacDougal Alley in New York City's Greenwich Village neighborhood illustrates the correspondence between material hardness and unconcealment (figure 9.9). In

9.9
MacDougal Alley, New York City, 1977 (photo by André Kertész).

the photograph, it is snow's transience and pliancy that is instrumental in revealing what harder and more permanent materials cannot.

Did the different material *Umwelts*, Matera and Inka, soft stone and hard stone, have any influence on their respective societies' social organization? Any claim of specific causality is speculative, given the innumerable factors and events underlying any societal structure. But to deny material a role would be to deny the mind's coupling with the *Umwelt*. The question I have is whether the hardness and permanence of the stone drove the Inkas to a command-and-control societal structure, since any ambitious construction with such hard material required a lot of highly organized manpower and time. And did the impermanence and softness of Matera's stone contribute to the small-scale, organic social ordering of its community?

The softness of the stone in Matera precluded the need for rigid hierarchical sociopolitical structures in order to build. In fact, building was a small-scale activity (as compared to that of the Inkas). Stones were cut to a regular—that is, uniform—size that two men could lift, which allowed for the family or extended family to construct in a prosaic manner. The Inkas, in contrast, had to muster enormous resources and exert a much more hierarchical order to build as they did. It is not inconceivable that the Inka ambition to construct meaningfully would not have taken hold in a landscape of loose, soft, weathering stone. The articulations of World, that of Matera and that of the Inka, were determined through the material they articulated, and the corresponding resistance of Earth was made present through those respective material properties.

Each society, situated and working in a different material *Umwelt*, brought forth its unique and particular *aletheia*—truth, unconcealment, and meaning.

Is it the same with modern society? Does material situate us in our larger urban and regional settings? What is the relationship between material and contemporary urbanism?

Cities seem to have far less connection to their material context. The brownstone of so many Manhattan residences and the underlying bedrock topography (which has influenced the height of buildings between Midtown and the Financial District) are more incidental than central to the cultural ethos of New York City. Ubiquitous glass towers and grids of streets and infrastructure overlay countless cities across the globe, giving a strong impression that the artifice of our built environment is no longer dependent on the materiality in which it is set. There is a growing sense that it is no longer a question of nature versus the built environment, but of the nature of the whole that encompasses planetary material, energy, flows, forces, and interactions, regardless of origin or whether organic or inorganic. A river's flow, a mountain of garbage, and a changing climate are equal participants in the anthropogenic era.

An interdisciplinary field of inquiry that emerged just over a decade ago, dubbed "new materialism," is reframing questions of the nature of material by removing the anthropocentric point of view and considering material's agency as part of a network of relations in which we are but another node or "actant"—"a source of action that can be either human or non-human; it is that which has efficacy, can *do* things, has sufficient coherence to make a difference, produce effects, alter the course of events."[17] In this view, material, far from being passively inert and awaiting mechanistic exploitation, has its own dynamism and organizational complexity and independent force.

The political theorists, philosophers, cultural theorists, and now even some artists and architects who have promulgated the new materialism reject traditional dualities of subject and object, active human and inert material. For them, it is events and emergent complex interactions, not isolated objects and entities, that frame material as deep and vital agents of terrestrial order. By dispensing with personal experience, the new materialists have uncoupled material from purpose and meaning.

As distant as such a "post-human-centered" condition may appear from the Inka and southern Italian peasant material-bound cultures, there is something held in common at the core: a belief that humans are but a part of the larger forces shaping the planet and that the physical *Umwelt* deeply penetrates the course of human events, even more so than we are aware. The affinities between new materialism and the Inka "correspondence with stone" suggest to me that material and its vitality were and remain key to our understanding of the greater cosmic order.

10 INSIDE THE DESIGN PROCESS

The discussion in the preceding chapter centered on the relationship between mind, material, and *Umwelt*, how material bears on experience and the mind, and how this might be reflected in cultures of building and making. As the ceaseless alteration of the contemporary world continues, design has assumed a greater role in shaping everyday life. How designers think about and make with materials, and how exactly material and design inform one another, bear examination, too—especially considering that designers and the design process direct the articulations of the materials that surround us.

Let's first situate the designer. Though they draw on the knowledge and abilities of engineers, makers, scientists, and poets, designers work with freedoms and within restrictions unique to their field of activity.

THE BRICOLEUR

Claude Lévi-Strauss popularized the term *bricoleur*, which he used in describing a modern-day equivalent to the mythological thinking or concrete science of earlier societies.[1] Loosely translated as a "jack-of-all-trades," the *bricoleur* putters about, working with the leftovers, with whatever is "at hand," constructing new arrangements from these former ends, now redeployed as means. The *bricoleur* picks up something and asks: What could this become? The *bricoleur* considers things as signs: halfway between concepts and images, signs can refer to concepts, but can stand as images as well.

Lévi-Strauss distinguished the *bricoleur* from the craftsman, the former's work being more "devious."

> There still exists among ourselves an activity which on the technical plane gives us quite a good understanding of what a science we prefer to call "prior" rather than "primitive" could have been on the plane of speculation. This is what is commonly called "bricolage" in French. In its old sense the verb "bricoler" applied to ball games and billiards, to hunting, shooting, and riding. It was however always used with reference to some extraneous movement: a ball rebounding, a dog straying, or a horse swerving from its direct course to avoid an obstacle. And in our own time the "bricoleur" is still someone who works with his hands and uses devious means compared to those of a craftsman.[2]

Lévi-Strauss also drew an explicit distinction between the *bricoleur* and the engineer, noting that "the engineer is always trying to make his way out of and go beyond the constraints imposed by a particular state of civilization while the 'bricoleur' by inclination or necessity always remains within them."

It is worth noting this even more strongly. Le Corbusier once wrote that the engineer is concerned with calculation, whereas the architect is concerned with organization[3]—underscoring the distinction drawn by Lévi-Strauss. Artists, architects, and designers can recognize themselves in the Lévi-Strauss description of the *bricoleur*. Rather than ponder the universe, they all address what might be called the "stuff in society's fridge." They are constrained by culture and limited to working within its implicit rules and language. They rely primarily on situated thinking, perception, and imagination to make relations, and tend to see ideas and images as coincident, in the form of signs. This is not to deny other means of working; the methods of science are also part of the thinking of an artist. But added to those methods are intuition, impulse, chance, reasoning, storytelling, deduction, induction, and magical thinking—all allowable operations for the *bricoleur*, and a larger set than that of the scientist; all appropriate for a project that creates relations where none have previously existed.

Unlike science, which is constantly attempting to go beyond the known, *bricoleurs* are bound by the things in front of them. They are forever rearranging the bits and pieces of culture, leaning in with the senses and on the lookout for meaning. The *bricoleur* works loosely, collecting tools and materials that are specific enough to be manipulated without highly specialized knowledge and equipment and that remain open to having more than one defined purpose or outcome.

"A particular cube of oak," to use the example employed by Lévi-Strauss, "could be a wedge to make up for the inadequate length of a plank of pine or it could be a pedestal—which would allow the grain and polish of the old wood to show to advantage."[4] In other words, the same piece of wood can change its meaning or purpose in an instant. There is no fixity, no "this is the proper way," no methodology. *Bricoleurs* thrive precisely because the new arrangements with which they work can be continually reimagined and rearranged to make their place in the world.

German psychologist Karl Duncker (1903–1940) coined the phrase "functional fixedness" to describe the mental tendency to block the reimagining of something so as to use it in a new way.[5] The cube of oak, for example, once understood as a wedge, will tend to get fixed in the mind as having only that function. Though functional fixedness is a trait of an object's perceived function, it might also apply to materials and media. I think this is why drawing with, say, a blue ballpoint pen in a notebook of lined paper can seem constraining: the paper, the lines, the pen are functionally fixed for the purpose of writing. This is also an "affordance"—the environment providing the opportunity for action—as defined by psychologist James Gibson (introduced in chapter 3). The bricoleur, paradoxically, is both on the lookout for affordances while escaping their accompanying fixities.

Duncker developed a cognitive performance test, still used today, to measure the influence of functional fixedness on problem-solving capabilities.[6] His "candle problem" involved providing subjects with a candle, a box of thumbtacks, and a book of matches and instructing them to attach the candle to a wall so as not to drip when lit. Most tried to melt wax or tack the candle directly to the wall; very few reimagined the function of the box by emptying it and tacking it to the wall, thus overcoming the functional fixedness of seeing the box only as the holder of the thumbtacks and instead designating it as a candleholder. Our *bricoleur* would likely have been happy to mess around with all the pieces, the "correct" arrangement being only one of several interesting possible outcomes.

Some might characterize this kind of nimbleness as merely "clever," intending to insinuate a facileness or a shallowness of intelligence. Serious people, all of them specialists in *something*, cringe at the thought of making do, of do-it-yourself, of the improperly trained "outsider," the jack-of-all-trades they imagine as some kind of caricature—perhaps a slightly disheveled, grinning character they do not know but to whom they assign all manner of eccentric traits. The territory of the *bricoleur* or mythical thinker is perilous, situated midway between perception and concepts. Although this territory lacks the authority of both the former's sensuousness and the latter's thoughtfulness, it has nevertheless been humankind's most fertile ground for discovery, invention, and creativity.

Most professions trace their origins to concrete science. For example, architecture—before it was a profession—consisted of builders, true *bricoleurs*, designing and building directly *in situ*. They had to adjust, invent, and "make do" with the situation at hand, adapting lessons learned from earlier generations. From the fifteenth century onward, though, architects formed a new professional class, gaining control of the building process, their drawings becoming the guiding instrument of construction. This laid the groundwork for a more "scientific" approach to design and building. With an almost exclusive focus on predictability and uniformity—which led to formalized training, licensing, and extensive codes and procedures—the architecture and construction professions have tended, over time, to devalue "concrete science." Part of this has meant the building and architecture professions neither celebrate nor even acknowledge the design process, which remains enigmatic and unknown to the public. The design process itself, however, has never abandoned the associations, narratives, metaphors, and mythological thinking of the *bricoleur*.

Serendipity is celebrated in science—the discovery of penicillin by Scottish biologist Alexander Fleming (1881–1955) being one such instance. He remarked "that's funny" when some unwanted mold contaminated a Petri dish in his untidy laboratory and he noticed that the mold had killed the surrounding bacteria cultures.[7] Design processes could be described similarly: the situation is almost always messy and too complex for methodology to digest; there's an active cultivation of chance and serendipity; there's a very broad allowance for things to "come to mind" and for a *bricoleur*'s proclivity to fool around with what is at hand. But while science celebrates its penicillin moments, the design professions—architecture in particular—tend to present their work as the product of pure reasoning, belying the fertile, even "unprofessional" design process behind the curtain. At most, the mythical "cocktail napkin sketch" represents a sanitized caricature of "process."

The project of science to develop structures based on theories and hypotheses turned away from the guidance of the senses and, in doing so, precipitated a rupture. As Kant put it, "Without sensibility no object would be given to us, without understanding no object would be thought. Thoughts without content are empty, intuitions without concepts are blind."[8] Said another way, thought not tied to sense results in the "thoughtless" or "senseless."

The real worth of concrete science lies in its fusion of thought and sense in the search for meaning and order in the world as it is found. While acknowledging the very real success and dominance of modern science, this "sensible reasoning"—found everywhere physical matter is touched—is the foundation of any design process. Even the surgeon performing a complex operation needs to sense, adjust, and work from whatever is at hand when something slips, splits, or turns out other than as planned. Anyone working with material, handling and manipulating physical matter, is inherently a *bricoleur*, flexible and able to work contingently and think with the senses, since material procedures never go exactly as intended.

Artists have a somewhat different relationship with their medium than designers. Paint to painters is both the means and the ends of their work, whereas materials for most designers are a means of work and the *specification* of an end—and more often than not, the two are very different. I admire the sense artists have of being in their medium, which so affects how they make. When a work of an artist is shown, exhibited, it is never an intermediate stage of something else that asks for our indulgence; rather, it is in the world "as is." Designers, though, are always making analogical pieces, using provisional materials, knowing full well that these are not the materials of the thing being designed. Architects are the most removed from the thing they are working on and have the most bifurcated material sensibility. Their medium/materials of design are completely detached from the materials of construction, so much so that it is common practice to think materially only toward the end of a design process. Artists have much to show designers in that respect—such as binding oneself to the medium in which one is working.

"How did you (or they) come up with that?" That is the question clients and lay audiences are always asking about the design for an object, a space, a building, or an environment. Just as when looking back on one's own life, looking back on a process tends to reinforce an orderly, linear view of events. In reality, however, the design process is far more lumpy than it is smooth, full of dead ends, mistakes, and quirky thinking. Its terrain is littered with debris. And since every design problem is a unique undertaking, it defies description. Like the stars in the night sky, the innumerable moments of a process might be recognized but not coherently. However, there are common topics that are part of every design process, and mapping them into constellations here can at least provide a glimpse into what lies behind design and why the science of the concrete is at home in the design process. There is no particular order; each topic remains in the present tense through any process; and the order of undertaking is contingent on the circumstances. There is neither beginning nor end, but only a constantly evolving situation that presents new obstacles and ways forward in the most unexpected ways.

WORDS THAT PROPEL

How a design problem is presented, defined, or framed is perhaps the most underrated inflection point in a design process. At the start of any design, there's a text, a brief, perhaps limits of time and money, a site, or a limit imposed by a fabrication process. Mostly, though, words—with all the hidden preconceptions and assumptions they carry—frame the situation. One might have a brief that asks for the design of an operable window of a certain size. That seems innocent enough: after all, who doesn't understand what a window is? But that is precisely what blocks the first design question: *What is a window?*

Etymology uncovers origins, often in the form of long-forgotten metaphors: *window* stems from the old Norse *vindr* ("wind") + *auga* ("eye"), which replaced Old English *eagþyrl* ("eye-hole") and *eagduru* ("eye-door"). Architecture professors are more apt to use "aperture" than "window," knowing full well that functional fixedness or fixity carries over into language—especially architectural terms. Once liberated from its fixity, "window" can be reconsidered as an aperture for light or air and/or looking, peeking, dreaming, and so on. The constant act of rephrasing and redefining terms, or problematizing, propels the design process and arms the *bricoleur* in the search for meaning and direction amongst the pieces with which she works.

Framing a problem in terms such as "design a window" implies that design is the process of solving that problem. Some problems require nothing more than a solution—how to hang a picture, for instance—but most, if not all, design problems are less a search for a solution than a search for *meaning*. Louis Kahn characterized the difference: "The creation of art is not the fulfillment of a need but the creation of a need. The world never needed Beethoven's Fifth Symphony until he created it. Now we could not live without it."[9]

Instead of framing "design a window" as a problem needing a solution, it can be treated as a search for questions worth asking—thus greatly expanding the possibilities. For a professional problem solver such as an engineer, though, such an approach would be unacceptable; indeed, it would be seen as an invitation to chaos. The engineer's instinctive response is to reframe the situation with enough limits, to establish sufficient parameters, so that there's a logical—that is, *computational*—path forward. Airplane designs are a good example of this: they have become designs of almost pure calculation, with visible design differences between aircraft vanishing as parameters are computationally optimized. In contrast, our window designer—like the mythological thinker—calmly embraces the proliferation of possibilities, searching for signs and relations.

Ultimately, every design is a story, with its mythologies and fictions, its facts and answers. No wonder, then, that Le Corbusier drew the distinction noted earlier: indeed, design leads with the *what* of things—and the story is the story of that meaning—much more than the *how* (the calculation of things). Problematizing enlarges the stage and makes room for a story to emerge that might never have been imagined.

MAKING AND BREAKING RULES

The 1979 film *Stalker* by Soviet director Andrei Tarkovsky (1932–1986) is set near and in the "Zone," a perilous, postapocalyptic exclusion area that can be entered and navigated only with the guidance of the Stalker, a mystic with special powers of insight.[10] He guides by throwing a steel nut and having his clients, the Professor and the Writer, follow precisely behind him as he moves from throw to throw. The path is convoluted: the shortest distance between two points is not a straight line, as all the rules of nature are different in the Zone. Eventually, the Writer ignores the Stalker's guidance and still manages to end up arriving at his destination, making the Stalker seem merely superstitious.

The story is an allegory of the mind and the frictions between *mythos* and *logos*. It is also an apt analogue of the nonlinear design process. As described in chapter 4, children will fill in gaps between events—even with made-up stories—in order to make sense of things. Working provisionally helps us proceed when the direction isn't clear. Like the Stalker's throws, provisional rules provide a trajectory; following that trajectory informs, and even if the result is a dead end, the information—which includes the trajectory—is valuable.

Rules, though, are always heuristic and never absolute, and one of the difficulties is to let go of rules after their provisional deployment. Even when they are successful in providing insights, they can quickly become impediments to furthering a design process—like a hammer when a screwdriver is needed. The window design, for example, might need some direction, and so I decide to limit all exploration to cutting and folding paper. I suspend judgment about this choice in the face of Kant's admonition that intuition without concepts is blind: if too much "why" is asked, the result is paralysis. After some time, I might realize the window is really a volume, a space, but one that now needs to be informed by vision (seeing in and seeing out). I am done using my working rule. It is no longer useful, and if I don't let go it will prevent me from working with the geometry of sight.

MATERIAL AS A MEDIUM

In a design process, materials are handy or useful to the degree they are easy to work. Why cut things out of stainless steel when cardboard of the same thickness is available? After all, it's a design, not the actual thing.

Material choices in such a mindset become more habitual, familiar, and prone to "material fixity"—a tendency to work a material within a limited set of habitual ways or affordances. There's plenty to reinforce this: tools, for instance, are a powerful definer of technique, and the design process itself is often oblivious to the role of materials. There is a broad range of attitudes toward material, from direct making—squeezing a lump of clay, for instance—to the immateriality of representing something already in "the mind's eye"—such as making a cardboard model of an object being designed.

If I choose to model the window with clay rather than cardboard, the clay will immediately assume an outsized role in the design process. Clay is a paradoxical material; again, as Bachelard wrote, it is "the perfect synthesis of resistance and malleability," "the imaginary paste" of the material imagination[11]—and yet it has a high degree of material fixity. I have observed countless students, when working with clay, fixating on their hand as the sole tool of manipulation, tactility trumping tactics.

With clay in particular, it takes time before other tools are considered, perhaps because of a strong belief that "the imaginary paste" can easily be bent to one's will. We are excited by clay in anticipation of its malleability, unaware of the subtle and mounting resistance to manual efforts that lies in store. If I try to squeeze and control the clay with my hands, to force it to behave as I might have planned, the initial clumsy formal outcomes can be quite frustrating.

Was the clay a "wrong" choice to work with, its reluctance to conform an impediment to getting *there*? Perhaps, but it will also make one aware of things that cardboard cannot. I might be inspired to control the clay and develop tools and mastery through craft, or I might switch materials, but regardless it will have compelled me to converse, to act, and to reason with it. This is the empathy discussed in chapter 7—a sensibility activated by the give and take, by the conversation between eye and hand, material and action.

Alternatively, I might choose to work with cardboard to represent the window about which I have an informal notion. The inherent planarity and cutting methods (drawn and cut lines) of the cardboard brings geometry inescapably into the foreground. Questions of dimension must be answered in order to make a cut and to coordinate the measured relationship between parts. Lines become drawing, and materiality recedes into the background. Not every material can be dematerialized in this way, but what the cardboard illustrates is how, by use, its materiality can be suppressed or altered. If I want to bring the cardboard's materiality into play, while keeping its drawing-like qualities, I could choose a much thinner stock—so thin that its floppiness would create all sorts of structural challenges and strategies (imagine making a beam out of tissue paper).

Let's return to the thicker cardboard I've drawn on, cut, and begun to glue together. As I proceed, my gluing is a bit messy, and in some places I might use tape to hold things together. All this is typical of study models that designers make, equivalent to a sketch or bricolage. What I find remarkable is that the materials—the cardboard, the glue, the tape, all of it—magically disappear from the design intent or thinking. Attention is focused on the sketched, formal aspects of the model and its eidetic qualities. All the rest is ignored; what it *represents* overshadows what it is *physically*.

In studio, we speak of "reading" a project as if it is all representation, a text as detached from its material as the morning's news is from the paper on which it is printed. I think it's safe to say that this is something learned, but in saying this we reveal how perception is a fusion of sense and thought and how much we look through stuff when we think we are looking at it.

As William Gass wrote, "Things have no meaning unless they stand for—become—signs."[12]

It should be noted that the transparency of representation depends on what is being represented. When one material represents another, such as the cardboard model of the window representing a metal frame it simulates, the representation is limited to "material transparency," wherein only the attributes of one material are read in the other; metal is read in the cardboard. In comparison, the reading of an actual construction—say, a brick wall—provides deeper readings: it can be read as representing the effort and skills of its makers; it can be read as being structurally solid (despite that nearly every brick wall is only a single brick deep, tied to the backing wall with "brick ties" and otherwise largely floating free); it can be read as an expression of honesty; and it can be read as a history of immigration and labor. It stands for something in addition to its material nature.

The spirit of the *bricoleur* in us is always on the lookout for signs, which are opaque and act materially yet can signify, can represent ideas and concepts. Signs, then, aren't necessarily present in the clay or chipboard sketch models, but were I to "see" in the facets of the cardboard or changing profile of the clay something *meaningful*, it would alter my perception. For example, if combining the tactility of the clay and the optical geometry inherent in the cardboard sparks a thought of a window that is formed from material plus language, it might lead to an anthropomorphic narrative of a window formed by fusing eye and hand. Perhaps the window is made of two materials: a tactile material wherever the window is touched and operated, and some other material where conforming to the ocular geometry.

An example of how meaning emerges in design can be seen in Sigurd Lewerentz's lighting design for his 1969 flower shop in Malmö, Sweden, at the entrance to a cemetery (figure 10.1). The humble shop, a raw, concrete, shed-roofed structure, defers to the mourners passing by and is devoid of any superfluous elements. Lewerentz chose to work with nothing more than what any hardware store carries: ceramic bases, bare bulbs, conduit, and so on. He designed bent and branched conduits emerging from junction boxes, "growing" lights along a wall like vines. But the logic

10.1
Lighting design, flower shop, Malmö, Sweden, Sigurd Lewerentz, 1969.

of their "growth" is inverted: plants grow toward light, but Lewerentz's electrical conduits grow toward the dark where the light is needed most. The design was careful to avoid the literal image of a plant, which would have diminished its meaning. Instead, the design remains abstract enough to be read/experienced as vinelike, as a yeoman's humble work, as construction for lygophilia, the love of darkness, and as a metaphorical antidote to mortality—seeking darkness in order to bring it light and life.

MATERIAL ANALOGIES

The role of materials in the design process can be understood as "opaque" or "transparent," which roughly correlates to the concepts of *entelechy* and *eidos* introduced in chapter 5. There is another role for material that is mysterious and surprising—that of an analogue and, as such, a lever for the imagination. Just as metaphors give form to the abstract with concrete terms, material analogues bring otherwise evanescent conditions into view. For designers, this is critical.

Material analogies combine the *eidos*-like, or eidetic, with the *entelechy*-like, or entelechic. Material is embraced and handled in its opaque and concrete fullness *and* is imagined as something other. This is not a problem of representation. The aim is more abstract and of the imagination: a meditation on possibilities, a search for relations, interdependences, orders, the emergence of things coming together meaningfully, before they can be named. Architecture has been called the most abstract of the arts in part because it is concerned with relations bound in constructed space; material analogues uncover abstract relations embedded in material behavior.

Here is an example. A few years ago, I gave a studio for students to develop a museum design based on the properties of material. They were asked to work with wood and, through actions on and with the wood, to find an architectural direction.

One student, Brandon Andow, had just returned from a trip to Japan. Inspired by the craftsmen he had encountered there, he began inserting small wedges into a simple slab of softwood, slowly expanding it without breakage. By the time he had reached a material limit, the wood piece had transformed from an inert piece into an organized and complex system of specific points, forces, actions, reactions, pressures, and limits. The value of this system was that it represented nothing; it was autonomous and open to interpretation. Although the work was physical, it evinced abstract qualities, triggered by the imagination: it was being imagined for what it could be.

Andow looked to his piece for clues to an architectural organization for a building, which required that he allow its scale to change. By imagining an 18-inch piece of wood as 300 feet long or as 80 feet high, he began to see spatial possibilities.

The paradox here is that Andow left the material, wood, behind in order to imagine it as an analogous model. The abandonment was not absolute; questions were directed back, and the original piece continued to serve as a laboratory of sorts. The wood wasn't so much studied for its properties as for how its traits could inform the development of the architectural order. Andow's model provided parameters and a set of variable relationships; it was more an adjustable instrument than a fixed configuration. Each wedge generated a gap, slid within limits, and effected adjacent wedges and gaps, which gave enough control for it to be a spatial organizing principle. Once the principle and rules were clear, a vocabulary or more specifically an architectural syntax slowly emerged—not a translation of form or wood properties, but analogous to something operative and addressed to the architectural problem at hand (figure 10.2).

10.2
"Wood Study," by Brandon Andow.

MATERIAL THINKING

Children draw water with a blue crayon, not because of any phenomenal insight but because blue represents water. We have seen that when material is the representation of thoughts, it (matter) is less generative—being bound to words and cultural codes makes it more like language and, like language, more stable. When material is worked with as a purely physical phenomenon, it will trigger the imagination and open possibilities—so long as there is a distance between the thing and what it has evoked, just like the distance required between the concrete and the abstraction to which it refers if a metaphor is to be effective.

The distance between Andow's piece of wood and his studies for an architectural order was sufficient to keep the dialogue open between the abstractions of geometry and space and the concreteness of wedges and bending, splitting wood. Eventually, Andow "mined" the wood piece to a limit for his project. For example, the laws of the curvatures of the stressed bending wood remained relevant, but also began to restrict the further development of the architectural order. There is a way forward from such fixity, which paradoxically is to leave behind the very material that birthed the original insights. If material is to act heuristically as a generator of ideas, it is a means and not the end; the ideas and relationships evinced by the material are what really matter (figure 10.3).

Let's say I crumple a piece of paper, intuitively seeking something, and find in the crumpling an image and a suggestion of space, topography, and order. I can keep crumpling the paper, shift to other kinds of paper, and learn a lot about the behavior of the material. But if the paper is the medium in which the project lives, it also obfuscates it. Say the pattern of ridges in the crumpled paper suggests something systematic (since paper crumples nonrepetitively but consistently); crumpling more paper will give

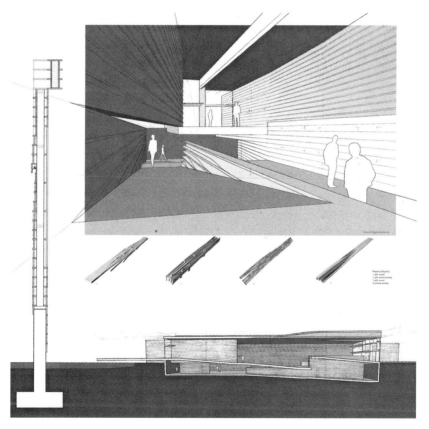

10.3
Architectural composition, Brandon Andow.

the same result. If I have to reimagine the behavior of the paper in a non-crumpling material, it will prompt me to bring reasoning to bear on the forces, reactions, and pattern of ridges at work in the paper. I will need to synthesize what occurred naturally and, in doing so, will have moved from questions raised by casual actions to questions of design.

As deeply as material shapes us and surrounds us, it has no inherent meaning. Wood itself does not *mean*. How, then, does material become meaningful in a design process? Time—and more specifically *lived* time—and experience over that time are what have the potential to give a material its meaning.

I still have a smooth stone I found on a beach in my childhood. It takes me back and marks my time and has meaning, if only for me. All the objects we surround ourselves with will pass into the past as they acquire history. When they no longer function, they become memories and, like religious relics, gain in significance and meaning with time. That is, the gap increases between their physicality and what they represent. A piece of wood might acquire meaning as it transforms from use or exposure or whatever interesting events it is subjected to, but this happens slowly over time, too slowly to be meaningful over the course of a design process.

How materials *mean* also changes with quantity. My colleague Robert O'Neal, who taught industrial design at Rhode Island School of Design for 34 years, used to give his students a simple assignment that illuminates this point. He asked students to design and make a cardboard box of a certain size. The next day, the same students were asked to design and make 10 boxes, and the following week to design and make a thousand boxes. When making one box, material was a conscious part of the students' designs, but when making a thousand their attention shifted to material as a conscious part of production. Their work became less about a dialogue with material and more about imposing logical procedures on material.

REDUCTIVE AND NARRATIVE APPROACHES

The examples I have given are snippets of design processes that have gone through many more iterations and material shifts on their way to a conclusion. This may imply that working from material properties calls for a reductive scientific approach in order to isolate and control salient aspects, to establish some parameters. It does to an extent. Equally relevant is the search for meaning, for things to be recognized as signs and to stand for something else.

Every mental decision of a design process is part *logos* and part *mythos*, and the nature of a design problem determines how they play out together. To use an extreme case, *logos* might be the only legitimized approach when designing an aircraft wing, but you can bet that intuition and arbitrary leaps—the stuff of *mythos*—have a role in the matter, albeit in the privacy of the designer's mind. The aircraft wing problem has a narrow goal, and the design process will advance from one iteration to the next depending only on performance outcomes. By contrast, the performance criteria in the design of a dinner plate have little influence on the design problem; the design process (hand forming and throwing the clay) is more likely to be guided by decisions based on material performance or a narrative context.

A reductive approach isolates relations and builds propositions based on concepts formed from these relations; a narrative approach accepts the whole and constructs a story to make sense of it all. That story always contains a part of the author, which reductive approaches instinctively reject in the name of reasoning. Each approach has its place. In combination, they can be at odds; if understood sympathetically, one can complement and propel the other. As Lévi-Strauss explained of the *bricoleur*, "he 'speaks' not only with things, as we have already seen, but also through the medium of things: giving an account of his personality and life by the choices he makes between the limited possibilities."[13]

When presenting work, especially early in the formative stages of a project, students often resort to a noncommittal narrative spoken in the passive voice: "This is what I did . . ." and "Later, what I decided to do was . . ." They substitute autobiography for an as yet unknown *logos* or *mythos*. The temptation to revert to habitual methods is palpable: "If only I knew how to proceed!" A design process falters if decisions are based exclusively on how to proceed, when the real search should be for the "what" of the project. The discomfort of not knowing, of being lost, underscores the emotional depth of our ordering instincts and the challenge of keeping any creative process "open"—without recourse to habitual method—long enough for things to percolate up from below, for the author to "speak" with them.

When we speak with materials, with things, we interrogate them with actions and listen with our thoughts. Narrative emerges from this conversation. A story arises in which we are but one character among others. The story gives voice to the inanimate, a voice mingled with our own, so much so that we can't be sure of our own authorship. A scientist might look aghast at all this cross-contamination, but it is exactly what creates fertile ground for the imagination. Narrative is a powerful way forward because it bridges unrelated events or things and turns the nonsensical into the knowable that addresses the "what."

OPERATING BETWEEN LOGOS AND MYTHOS

The design process shifts constantly as one's work presents itself in images from which concepts come to mind. Signs, like those floating bits in our eyeballs, mediate transparently between what we perceive and what we think, being neither but able to accommodate both and thus playing a crucial role. Signs function by association: an unknown mark or thing gains meaning when associated with something that is meaningful. That is the primary unit of narrative and representation, just as a word is to language. In *The Savage Mind*, Lévi-Strauss gives an example of a woodpecker's beak resembling a human tooth, fostering a narrative of touching the beak as a cure for a toothache. The narrative gives order to the relationship between tooth and beak, addressing truth through the lens of meaning.

For narrative to propel a work, we must be on the lookout for meaning. Everything must be considered a potential sign, while we keep the primacy of perception intact. If we impose preconceptions on the work, it will lose its ability to speak, to guide us.

One of the material-based studios I taught was centered on the casting process. The class was asked to work with plaster as a generative material, as a medium of spatial exploration. The work of Tae Park, a student in the class, illustrates how narrative and reductive thinking can combine to empower the design process.

The architectural problem was simple: develop a chapel through exploring a casting-based process. Short assignments gave the students some structure. For instance, I asked them to develop a cast wall with openings for light. Park's wall studies weren't particularly novel, or well crafted. They exhibited typical flaws of a novice cast, including "eroded," dendritic-like surfaces where water had separated from the setting plaster. At this stage of a design process, such an unforeseen result is typically disregarded as an error, distracting from intention and having no value because it hadn't been sought. But Park took note of this unanticipated trait—not as a failure, but as something significant. It triggered a thought of a branching and rooted tree as a metaphor for the connection between our grounded self and the "all" above (figure 10.4).

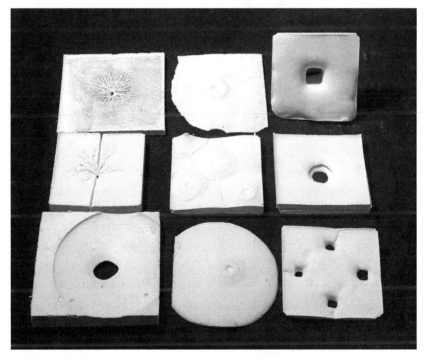

10.4
Initial experiments in plaster, Tae Park.

As his project developed over three months, Park's early intuition proved pivotal. It was hardly blind intuition "without concepts," as Kant put it,[14] nor was it only an image. The fissures in the plaster, the material play between liquid and solid, fragile and strong, filling and eroding, were sensed as signifying qualities—but not the appearance—of an architecture to come.

As part of the studio's pedagogy, I gave the students another assignment: make a vessel within a vessel. I asked them to work this through, first considering how they might achieve a neutral (uninflected) condition with one vessel nested within another, and then transforming the relationship between the two into something engaged, inflected, embracing, reciprocal, and so on. Park began spinning a simple box container (on the end of a drill) and poured plaster in as it turned. At first, he controlled the form, sweeping a radius to make a cylindrical space in the container's center. He soon realized that control needed to be shared with the material process, and so he began centrifuging the plaster at a high enough speed to have it climb the walls of the container, the result of which was to form an organic cone. To transform the neutral relationship, Park began to move the center of rotation, resulting in plaster sloshing unevenly up the walls of the container, which created eccentric voids.

And on it went, a kind of miniature laboratory with procedures, measures, and hypotheses. The last step in this series of logical steps was to cast another space within the first result, to create vessel forms formed by physical forces both inside and out (figure 10.5).

Like a true *bricoleur*, Park had nimbly adjusted his apperceptive process from the *mythos* of the tree to the *logos* of the plaster vessels.

His was already a rich harvest, but Park didn't stop. He couldn't really, because the project had reached a point at which it asserted its own inertia that had arisen from the symbiosis between the empirical studies and a nascent narrative arc. Using the inner vessel as a formwork, Park constructed a stepping and rising architectonic (fundamental parts of architecture) figure, not yet architectural but beginning to express affinities independent of the plaster medium. As is characteristic of narrative, language also returned as a medium to give definition to his work. "Gravity,"

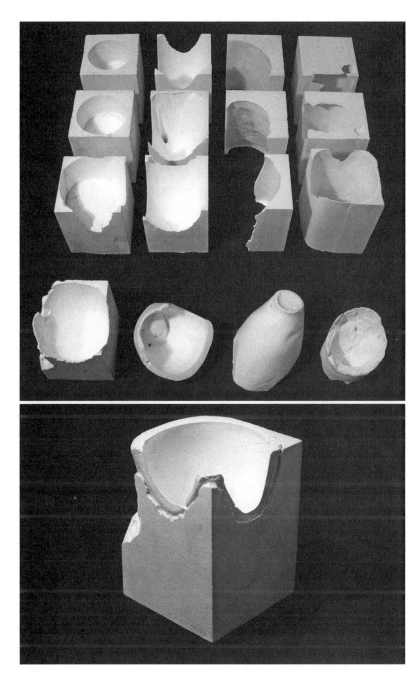

10.5
Plaster vessels, Tae Park.

"ascension," "light," "weight," "spinning eccentric bodies," "sinking," tree of life," "rootedness," and "reaching" all began to coalesce as a concept and a story. The parabolic intersections of the "pure" vessel with the orthogonal container became "aperture," and the plaster vessel itself was reconceived as two types of elements, one solitary for individuals and the other as a collection for a gathering. Slowly, the narrative and the material logic wove together to catalyze a design. Park had designed a chapel.

MODELS

The word "model" is rooted etymologically in the Latin word *modulus*, meaning "a small measure," which itself is a diminutive of the Latin *modus*, meaning "manner," that found its way through Italian to *modello*, meaning "a model" or "a mold." Over time, the noun also became a verb.

In architectural models, what is being measured and in what manner vary widely. During the first stages of a work, designers directly address the things they have created and speak privately with them, arranging, editing, and remaking the work. At that early stage, models are tries, instruments, attempts to render visible. They are under no obligation to conform to convention and are like a writer's notes.

At the next stage, however, a representational model is pivotal, turning the work outward—that is, away from the author toward an audience—and acknowledging its social dimension. Fine artists can put off this moment until after their work is complete—such as when they exhibit in a gallery. For designers and architects, the representational hinge point is embedded in the heart of the design process, framing the work in its cultural context and remaking it within the limits, values, and terms of society (often misconstrued as "reality"). Representational models are signs inasmuch as they "stand for" something. Representation pivots material, transforming it from *being* to *meaning,* from actuality to language. For example, clay—which is used as a medium of exploration, to be squeezed and formed—can also (depending on the scale of representation) represent another material, such as masonry or concrete, or the solidity and mass of a building.

This may seem confusing to the nondesigner. Aren't models just little miniatures of something? When beginning design students make things in studios, they tend to think of a model as a small copy of reality and aim to imitate each element with a material that "looks like it." This is like using random words without the syntax required to construct a meaningful sentence; it leads to lots of things, with little or no relationship among them. Such novice models are opaque; they neither participate in the design process nor articulate concepts. "Looks like it" displaces "be like it."

(It should be noted that there is a role for sophisticated versions of such models in design fields. Long after any design process is completed, such models, including rendered digital models, are typically made for marketing purposes, to appeal to eyes with an appetite for the spectacular, which might in part explain the attraction to lay audiences.)

As Park's design evolved from its material and conceptual expression, material began to shift from being considered for what it was to something more representational. An early study model of his chapel building (figure 10.6) is a hybrid, both representing the building and actualizing another material process: coiled string constructs the individual light shafts; developed surfaces define the gathering light elements; and simple masking tape wraps the chapel to form a swelling body tethered to the party wall of the site.

In figure 10.6, we can see both the material—the *entelechy*, the actualization of forms from forces—and *through* the materials to the ideas represented—the *eidos*. The implied movements of each technique—coiling, wrapping, seeking, falling, and wobbling—have constructed meaning and meaningfully constructed the chapel.

Park's "final" model (figure 10.7) shifts the material palette to that of conventional model-making materials, in this case chipboard and card stock (with the exception of the string). One purpose of such models is to test propositions; in the case of architecture, models are (almost) always at a scale, which like projection drawing dematerializes its subject. As Louis Kahn wrote, "A great building must begin with the unmeasurable, must go through measurable means when it is being designed and, in the end, must be unmeasurable."[15]

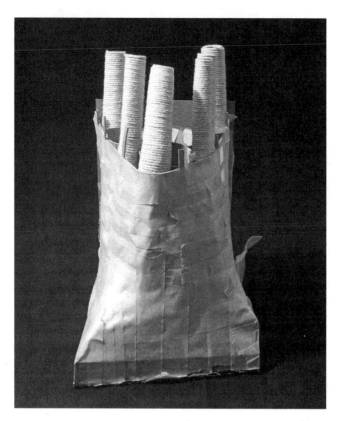

10.6
Study model of chapel, Tae Park.

Park's model is another stage in the progression to the measurable—
the eye of the needle through which all things built by others must pass.
As with projection drawing, scale models—despite appearances and having
inherent qualities, especially material qualities—are poor at representing
qualitative aspects of a proposition. Traits of geometry, dimension, form,
spatial relationships, all scale up or down seamlessly, but phenomenally
grasped aspects such as color and material are a different story. For instance,
we perceive wood as wood no matter what other form it may take, including as an architectural model. When we use wood to make an architectural model at scale, as is typical, that wood is not conceived as being part
of what the model represents as a constructed proposal (that is, we don't

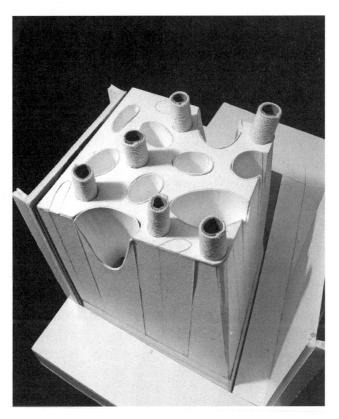

10.7
Final model of chapel, Tae Park.

assume that the ultimate building will be made of wood). The wood is an abstraction: a wooden model of a concert hall, for example, is read as *just* a concert hall and not a *wooden* concert hall. In other words, a wooden architectural model is understood as immaterial despite its materiality. Indeed, wood's materiality changes once it is used as a part of a scaled representation such as an architectural model; the wood of the model recedes to the background, seeming to dematerialize itself.

Another value of Park's model, besides allowing us to imagine with and through it, is what it elicited from him. Everything we make puts demands on us, although these demands tend to sneak up on us in the guise of our own volition. Take, for example, the final model of his chapel (figure 10.7).

Much of the information necessary to make the surface of the model was easily transferred directly from drawings. The cones (which in the model in figure 10.6 had first been developed by unrolling paper cones) required a number of additional steps, including pattern making, the design and layout of facets, and cutting and assembling the parts.

The "terms of making" called for by the final model demanded architectonic clarity and a higher degree of precision than the previous models (figures 10.4, 10.5, and 10.6). To some degree, the immateriality of the last model helps focus attention on geometry and tectonics—the logic of parts—and, in some respects, marks the first, albeit abstract, meeting of architecture and construction. The wrapped string tubes are an exception, having remained unchanged from previous iterations, not because they had reached some kind of resolution but because their actual constructive logic remained elusive.

INCREMENTAL CREATIVITY

The spark of discovery or invention or creativity doesn't jump far. Kepler, known for his incredible leaps of the mind, in fact made tiny connections that had enormous consequences. His discovery of the elliptical orbit for Mars detailed in chapter 5 was really the recognition that one number was the same as another. In and of itself, that was not a huge leap, but Kepler's realization of the *meaning* of it made it an enormous breakthrough for humankind. And so it goes, whether we are talking about the history of scientific discovery or a design process: all steps are incremental and small. The development of the integrated circuit is a typical example. Over many years, many scientists contributed a new part or some improvement, without ever a great leap.

Park's project was also a series of steps, and never a leap. The canard "leap of the imagination" has us believe in "ahas" that just pop into our mind without preparatory work. The relative rarity of creativity isn't due an inability to make imaginative leaps, but because it takes a great deal of effort to bring two unrelated things close enough together to make a leap. The design process is similar to building a tunnel from both ends blindly:

nothing is in sight or seemingly gained, despite a lot of work, until there is a literal breakthrough—which takes only the last shovelful (of imagination) to accomplish.

WORKING DRAWINGS

How is a design translated from a sketch or scraps into a precise and finished thing? The answer shows further how material is affected by thinking and how, in turn, it affects thought. It also reveals the effects of drawings on material, one of which is to make materials drawing-like.

The process is a bit hermetic to anyone unfamiliar with the technical drawings used by engineers, architects, and designers. "Blueprints" or "working drawings" are full of ciphers, symbols, numbers, and odd shorthand such as T.O.S and T.O.F.F. and R.O. (which, by the way, refer to top of slab, top of finished floor, and rough opening, respectively), giving the strong impression that only those in the profession know the code to create and read such documents. It seems working drawings have little impact on the design process—they appear after it is finished—or even on the thing being made from the drawings. The very term "working drawings" implies that this delineated information is just about doing the work of mules, moving something from here to there in yeoman-like fashion, all technique.

In reality, the progenitors of working drawings are used throughout the design process: plans, sections, elevations in architecture; cuts, views, and cross sections in industrial design; and patterns in apparel design. Most of these drawings are orthographic projections, in that all the lines of projection are parallel. For example, an orthogonal drawing of the side of a box will preserve its dimensional information no matter how far away from the box the drawing is made. This allows the drawing to be used to fabricate the side of the box accurately. Not only are such drawings a fundamental activity of designing, but they also influence the design process and the materialization of design. In other words, the spaces and objects we design are influenced by the very medium that carries the information to make them.

Let's say we want to design a box. A box's six faces correspond to six views of it: top, bottom, each end, and each side. We can create a blueprint that shows all six views and use it to instruct a friend who will cut out six faces of material and assemble as directed. That seems simple enough. We can even elaborate on the design by drawing openings on any or all the faces as we wish, and have our friend cut out those openings, without error or loss of accuracy. As long as we keep the relationship between the drawing and the thing constant, it seems our blueprints are doing fine, just propagating the information necessary to make the box. But try changing one side of the box—say, to make it curved—and suddenly the flat drawing cannot correspond to the curved surface. Making that new box becomes exponentially more difficult.

Hiding in plain sight is the powerful influence of drawing to flatten and make things boxlike simply because drawing anything that isn't planar and boxlike and aligned with the x, y, and z axes becomes extremely challenging to handle within the drawing system. With today's digital tools, wresting forms from surfaces is no longer as difficult; however, xyz spatial conventions still rule the virtual space of computer models, paralleling the restrictions of orthographic drawing.

The logic of orthogonal projection comports with the world of xyz spatial conventions we have built for ourselves, which is a fundamental reason that buildings are, for the most part, boxes. This marriage of convenience allows buildings to be drawn and built easily. Even within the design process, drawing on a flat piece of paper tends to reinforce and generate a conception of a flat surface.

Try walking about with this in mind, and you will begin to see that the vast majority of buildings and their interior spaces are, in essence, drawings, flat and orthogonal. It is rather uncanny to see blueprints surrounding us, and yet they are very much part of our everyday environment. Also notable is how materials are pressed into this flattened world. It is not accidental that materials are manufactured into thin sheets, and that construction consists mostly of building up many layers of flattened materials. Given this flattening of materials, it is small wonder that materiality is even more elusive in the design/drawing stage. If we are designing a box, at what point do we consider its materiality? Our working drawing cannot "feel" the material differences between cardboard or wood or steel; this gap between mind and material is only overcome with tacit sensibility, including material empathy, physical experimentation, and experience. Those who make understand this.

Given the prospect of a rather boxed-in and immaterial approach to design and architecture, it's no surprise that many students feel the urge to bypass drawing and build directly. And why not? The hand connects directly with the work, and there is immediate feedback and strong sensibility gained from the physicality of it all. But this freedom of action comes at a cost. Conceptual and organizational richness and the development of complex relations between elements, space, and human experience depend on an exquisitely balanced ballet among space, structure, geometry, form, inside, outside, light, and earth. A project is by definition an anticipatory work, and projection drawing is (still) unmatched in its capacity to foresee and plan free of the ponderous irreversibility and opacity of physical construction. The very immateriality of drawing can be a powerful instrument of insight.

Architectural drawings are deconstructions, each an isolated fragment of the totality of a thing. Every drawing can preserve certain information (measure *or* appearance, but never both). Each fragment/drawing is autonomous—subject to its own rules, forces, needs, cultural baggage, and language. Take, for example, a plan drawing. It constitutes a generative plane that will never be seen or experienced as such. Its generative power derives from its limits and simplification to two dimensions, which allow one to "see" the relation between empty and solid, inside and outside, order and inhabitation—sublimating experience into the density of organizational geometry. A plan is a good instrument because it is graspable and controllable—it can be readily steered and adjusted by insights—and, like a passing cloud, it is just simple enough for the imagination to flourish. Likewise, a section drawing (a projection of a vertical slice) brings forth the experience of physical weight as an upright body does (by virtue of being parallel to gravity). It is in section drawings that the presence of structure is first felt, and the designer begins to see how that structure might provide shelter, how water might run off, and so on. Each drawing, by virtue of its selectivity, can enter into a reciprocal relationship with the author: a mark is made, its impact noted, leading to further acts.

At the cost of losing direct contact with material, the various orthogonal drawings such as I have described constitute ways of abstract thinking and imagining that have empowered architects and designers for centuries, albeit with the (hidden) restrictions and tradeoffs I have discussed.

An alternative approach is to bypass drawing and work in model. Here, too, there are tradeoffs. Whether a model is physical or digital, it inverts the difficulties of drawing: the whole is no longer a furtive and abstract synthesis of multiple drawings, but rather its plans and sections are after the fact, derivative cuts, emptied of their power to author and guide the work. Herein lies a continuing paradox of the design process: orthographic plans and sections are crucibles of content and meaning and models are harbors of form. Since form without content is meaningless and content without form inchoate, it is the lines of correspondences between orthogonal drawings and between drawings and model that give content its form and that transmit the immeasurable.

Too little attention has been given to this issue of content's relationship to form and the role of drawing. It is either treated as a problem of technique, as a how-to, or—in the digital arena—simply ignored.

SCALE

Drawing an image of a figure or sketching a design idea is reductive because the object of inquiry has shrunk. This reduction of complexity is powerful because it unifies and allows us to grasp the whole. When a thing's size is diminished, that thing becomes abstracted, and our own empathic, tactile sense of it also diminishes. Who hasn't looked out an airplane window while flying and become more attuned to the daydreamt thoughts flowing inside one's mind than to the distant ground below? Historically, painting almost always involved a reduction in size, whether it was a landscape or portraiture, while traditionally sculpture has remained close to "full size." Extreme enlargement or diminishment in both the former and latter disciplines pose technical challenges more than conceptual ones; for this reason, miniatures and enlargements in both painting and sculpture are well established and wholly within the framework of their respective disciplines.

The design process in architecture is much more scale-intensive, involving a lot of scaled drawings and models. It is important to note, however, that sketching and working with materials are also done as "full-size" activities as part of an architectural design process. I make this point to address a common misconception about the design process, especially with respect to architecture: that, since design is done at scale, it must exclusively be an activity of making representations. This, unfortunately, leads to another misconception (the one that motivated this book and its exploration): that we first think, and then represent our thoughts. Even architects often fail to see the influence of ways of working on the thought process, because so much of our attention is devoted to working within scaled drawings and models. What is not widely understood is how the architectural design process employs scale as a way to induce thinking.

Of the visual arts, architecture has the distinction of doubly operating at scale, first in the sketches and models of a design (which are dramatic diminutions of reality) and then when the products of a design (scaled representations such as drawings and models) are massively enlarged to "full size" in the construction process. The architect goes through a double scaling of reality, down and then back up, doubling the transformative ramifications. Scaling up and down is also an operational part of the design process. For example, by reducing the scale of a design drawing or model, the "whole" can be induced. The "thumbnail sketch" (so named to emphasize the small amount of detail a thumbnail-sized drawing can carry) diminishes the details of a design, limiting (and empowering) the hand/ mind to focus on essential elements and relations. This field of operations is unlike miniaturization, which tries to preserve as much detail as possible. Simply consider the classic example of a ship in a bottle.

The thumbnail sketch is a point of origin, generative, mysterious, and pregnant with meaning, where a mark or element is born and—by virtue of its abstraction—can bridge between image and concept. This is more difficult at larger scales, which have fixed relationships and an increasing focus on the articulation of and connections between elements.

Consider the thumbnail sketches in figure 10.8. By offering clues and possibilities rather than an explicit representation, they trigger an internal feeling—which might be called joyful or empathic imagination. It's easy and exciting to imagine the possibility of one of these being a building because of the immense difference in scale. This incompleteness and distance, if calibrated just so, can catalyze image, concept, sign, spatial order, and experiential order. Too few clues, and the sketch ends up being no more than the sum of its parts, too many and it leaves nothing to the imagination. Empathy, which normally falls away with distance and abstraction, is here difficult to differentiate from imagination, providing an important clue about their intertwined nature: an image can only be evoked when we "feel into" a thumbnail sketch.

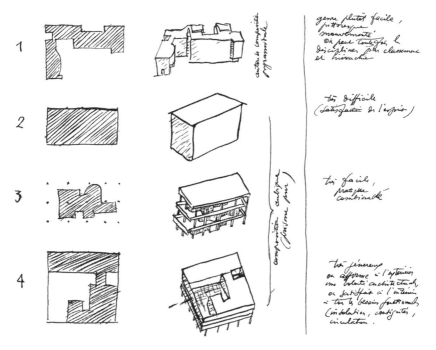

10.8

Thumbnail sketches of four house compositions, Le Corbusier, 1929.

A thumbnail sketch may seem diagrammatic, but it isn't a diagram. The difference between thumbnail sketches and diagrams is easy to overlook since they resemble each other: both are small-scale, immaterial, and constituted of distinct elements. Design and architecture use diagrams to reduce designs to a visual logic, mostly as a drawing of geometric elements and/or fields. Diagrams are analytic tools, akin to a lens with which to see more, and as distinct from a thumbnail sketch as a chemical equation is from DNA. Diagrams are reductive, preserving the metrics and geometric facts of elemental parts from which rules of assemblage and syntax can be constructed. Like a scientific hypothesis, diagramming is predicated on the idea that logic is the sole organizing authority of a project.

The "thumbnail" is a distillation or condensation. Like narrative, it tries to hold the whole in some kind of meaningful order—the purpose being to establish consequential relations, not by mechanical operations such as overlaying diagrams but through composition, ordering, and organizing. There is an affinity between the thumbnail sketch and material-led design processes; both are irreducible to information and data, and both, by their nature, resist autonomous juxtapositions. Everything is consequential in materials and thumbnail drawings, which mediate three-dimensional relations and orders. In short, they mediate the *whole*.

As with material, the thumbnail also takes time and sensitivity to master. The earthen fecundity of a thumbnail stems from its potential to blossom into narrative. Thumbnail sketches are founded on meaning evinced by elements and their relationships. The limited attributes under consideration, including materiality or even a consistent scale, help resist the easy image, the quick filling in of the blanks.

While it is counterintuitive to think of a paucity of stimuli as inspiring creativity, we know we see more imaginatively when gazing at a cloud than at a screen. This inspiration by "lack of" can be observed in an entire culture; Italy, for example, facing limited food diversity and resources through much of its history, still developed one of the world's most inventive cuisines based on a few humble ingredients.

Scaling down clarifies through consolidation and unification of the whole. Scaling up, by contrast, is downright bewildering. Within the design process, simply doubling the scale of work, even on paper, calls attention to all that is missing and provokes a sense of emptiness. As scale increases, the lack of materiality and physicality is acutely felt; unlike the triggering density of a diagram or thumbnail sketch, picturing an enlarged scale is like trying to look over a horizon—quite a difficult task, even for the imagination.

Our capacity to anticipate and feel for the consequences of scaling up is very limited. Take, for example, the relationship between length and volume. Our natural inclination is to extrapolate scale linearly. We don't instantly perceive that a doubling in scale results in a fourfold increase in area and an eightfold increase in volume (and mass). The contrast between the slenderness of mice bones and thickly proportioned elephant bones illustrates this proportional fact superbly. Or consider the weight of a car at the scale of a toy. The famous and immensely popular Dinky toy cars from England (which predated Corgi, Matchbox, and Hot Wheels cars) were die-cast in metal at 1:45 scale. A fifteen-foot car at that scale would be about 4 inches long. I remember their indestructible toughness and heft— they weighed around 5 or 6 ounces. Scaled down in terms of mass, the toys should have weighed only about a half-ounce. That means they would have to have been made from one-tenth the amount of metal. Imagine how fragile that would have made the toy. Conversely, scaling up the mass of the toy following the 1:45 scale would have resulted in a full-size model weighing a hefty 40,000 pounds or so.

Engineers, armed with their rules and formulas, are the seers in the scaled-up world we call reality—scaled up because so much of what we make originates on the drawing board at a far smaller size. Tiny, imperceptible material properties are magnified when we make giant things, revealing hitherto unimaginable behavior. Our assumptions of stability in our surroundings are off. Everything is in flux, and we often don't notice until it's the proverbial "too late." The Verrazano-Narrows Bridge that connects the New York City boroughs of Brooklyn and Staten Island, for example, rises and falls *12 feet* over the course of each year just from the influence of temperature on the steel cables. Had that not been factored into the design of the bridge, it would have self-destructed.

When we consider the future, the most powerful tool at our disposal is projection—extrapolating the present forward in time. The same with scaling up: we tend to extrapolate from the scale at which we are working. And as evidenced by the challenge of relating the scaling of dimension to mass, scaling up quantifiable aspects is difficult enough. In the process of scaling the immeasurable, the *qualia*, things get exponentially more challenging and transformational, undergoing a sort of capriciousness with unpredictable consequences.

When I was a young architect, just out of school, I was assigned the task of supervising the fabrication of a large-scale model of a high-rise building my firm had designed for a developer. The fabricator, an old hand at this kind of work who had made hundreds of such models, had a few questions that were easy enough to answer—except one. George asked, "What color do you want me to make the brick?"

I was stymied. I knew the actual brick color, but had no idea how to scale it to represent the color at 1/200th of its actual size. My roommate, who happened to be a colorist for a textile design company, offered to help solve the problem, and so we set about mixing colors—lots of them—in an effort to create a palette for the model. Even with a solid understanding of color theory, it took quite some time and numerous failed attempts before we made any headway. What might seem to be a sensible approach, such as sampling colors from images taken of bricks at a distance, was ineffective. So, too, was working with color theory, such as blue shifting to account for distance. (The color blue in the visible spectrum has a shorter wavelength than most other colors; in photochemistry, a change of spectral band position in the absorption, reflectance, transmittance, or emission spectrum of a molecule to a shorter wavelength is commonly called a blue shift.) What worked best was a counterintuitive palette that included gray blues, lavenders, and pinkish and greenish hues. Each color was wrong, but as a set the palette best represented the materiality of the scale model—not unlike the way an impressionist painting changes its chromatic character as it is viewed from various distances.

Juhani Pallasmaa, the Finnish architect and writer, has quite notably decried the hegemony of the ocular in architecture and the displacement of the other senses by visual spectacle.[16] What he does not discuss is how scale affects ocularity. As we design things that are increasingly scaled up from what we can experience, the senses fall away, quickly leaving the visual as the predominant experiential sense. The larger the structure being designed, the greater the scale shift between what's on screen or on the table and what will actually be constructed, the more the ocular comes to dominate the design process.

There are other important agents at work that suppress experience. Most design work, including architecture, is consumed on the page or screen, and the explosion of publications has surely biased the evaluation of work toward the flatness of the image. Further, the ubiquity of digital tools since the late 1990s has profoundly changed the design experience from a tactile and physical undertaking to an almost exclusively visual one, paralleling the ocular domination of architectural experience. An exploration of these tools, both physical and digital, is warranted.

11 DESIGN TOOLS AND THEIR ROLES

Artists' art stems from the unique choices they make in the ways they work, unique because of each artist's particular, individual sensibility. Tools in this respect might be regarded as extensions of the will, chosen and manipulated freely, or conversely as so influential on the artist's/designer's process that much of any creative outcome is steered by the particular choices of tools and materials. Tools in the former view are transmitters of artistic sensibility; in the latter, they inform artistic sensibility.

Any discussion of the tools of design assumes artists and designers choose tools based on their particular inclinations. The French poet and philosopher Paul Valéry (1871–1945) observed that we tend to blur two characteristics of artists: one, their innate talent; the other, their acquired skills and experience in the means of execution. He defined the relation of the two characteristics like this: "Art, in this sense, is that quality of the *way of doing* (whatever its object may be) which is due to *dissimilarity in the modes of operation* and hence in the results—arising from the *dissimilarity of the agents*."[1]

Just as the brush is associated with the painter and the chisel with the sculptor, the pencil is associated with the designer. But these are also clichés, and clichés promulgate bias. In the case of the pencil cliché, that bias is the perception that design is primarily an activity of depiction, with varying degrees of precision ranging from sketch to blueprint. There is some truth to this. A pencil makes lines, and lines can be measured—which foreshadows the inevitable demand on all design for reproducibility. Associating the pencil with design also reinforces a persistent perception of design as a more cerebral and somehow less "artistic" process, the designer first pondering and absorbing information and only then sketching out a design in an almost antiseptic, dematerialized process. In truth, design preconceptions limit the design process more than do tools, media, or materials. The strongest of these preconceptions is that the design process is a matter of visualization, of getting stuff from one's imagination onto the page.

About fifteen years ago, Dietrich Neumann—an architectural historian and colleague of mine from Brown University—gave a lecture on virtual reality and specifically on how that technology and emergent visual modeling software might revolutionize architecture—but not necessarily for the better. Just as small-format photography had created a huge class of amateur photographers, he posited that visualization software would put the means of design in the hands of everyman. A lot has changed since that lecture: design software has become more accessible, simpler and even freely available, but the design field is hardly being overrun by amateurs seizing on this massive democratization of digital tools. What, then, is holding everyone back? Just as access to a typewriter cannot transform anyone into an actual writer of literature, access to digital software—3D software in particular—does not make anyone a designer. Were visualization the problem, my colleague's prediction would have already come true.

I wrote in chapter 4 of a drawing experiment in which I had my students draw an apple they imagined, and how they consistently ended up drawing unimaginative, cartoon-like apples. One *might* conclude that the students lacked visualization skills; however, ten years of watching some of the most visually gifted students from around the world draw anemic apples makes it clear to me that visualization is not the issue. These drawings all indicate that the imagination is as much or more "out there" (in the world) as "in here" (in the head), and that it depends on and grows with our perception. The "in here" imagination of the students is a preconception. In other words, the humble apple on the table in front of us has more to it than can be imagined. Drawing that apple employs both memory and imagination to bridge the temporal and spatial gap between observation and mark making.

To draw what we see requires us to remember what we are looking at as we make a mark on the page. Our imagination allows us to nudge that mark further than what we just saw. The disconnection between seeing and marking isn't mere nuisance; rather, it is essential to the creativity and the authorship of drawing. In that gap, our own particular "vision" emerges—propelled by the misalignment of the subject and its representation, perception, and preconception. It is a misalignment that vitalizes the traffic back and forth between apple, eye, hand, marks, and the imagination. The "imagined" apple drawings lack this gap; what is conjured by the mind is transmitted to the hand as a trace, a command, with little or no reason to recruit the imagination to the act of drawing.

I am reminded of Merlin Donald's three types of visual memory discussed in chapter 2: the episodic, procedural, and semantic.[2] Could the procedural (the type that stores only generalities) overshadow the episodic when tracing out the imagined, referencing only a generic image? I am also reminded of how much the act of drawing is a perfect example of the enacted mind. Robust visualization is a two-way street: what is envisioned in the mind depends on what is seen, which renders the relationship between the imagination and design mutable and unstable. If for no other reason, design tools and materials—agents in this turbulent back-and-forth—have a deep influence on the design process.

This influence goes unnoticed. As we learn about and practice with material, media, and tools, they increasingly pull and align our intent and method to their limits and affordances. This alliance—one often referred to as the mark of mastery—is cloaked by its positive rewards; the lurching missteps of a novice offer none of the satisfactions that come with graceful proficiency. And since mastery of this sort involves tacit knowledge to which we have no conscious access, the effects of the tool on the master's imaginative range are neither known nor felt—that is, unless this alliance is disrupted.

Were I to initiate a design process by choosing not the default pencil but instead an unconventional tool, would my choice affect the design? Using a table saw at the formative stage of design would immediately ensure that any outcome would be tied to the saw's limits and capacities.

The emergent design language would be planar, chopped, slit, ripped, angled, and parallel. It would not and could not involve much else, and the ensuing criticism would be directed at the way in which the chosen tool limited and controlled the creative process—points that would never be raised had a pencil been used.

Even the paintbrush can be a problematic design tool if one is habituated to working with pencil or pen. The "problem" of the paintbrush is its culture. To be sure, there is an enormous amount of technique in controlling the brush—a lot of knowhow—but the freedom it offers to splatter, smear, and the like makes it an unlikely tool for the linear depiction afforded by the pencil.

What we can glean from thinking of the saw and the brush is that the pen and pencil are more strongly linked to depiction than we may realize. For the Greeks, *graphe* is a root common to words meaning both writing and drawing, and I suspect the propensity to use line drawing tools is related to this: pencils are used in design in a similar mental fashion to the way words are written—that is, marks are placed on the page as visual interpretations of thought, not as nonrepresentational abstract marks.

My point is that as long as tools are used "properly," the surface of consciousness remains undisturbed. Each tool is subtly held within a conceptual framework that becomes only apparent when that tool is misused or misapplied—say, when paint is applied with a wire brush. That alters and erodes the plane of consciousness; something is exposed, revealed, that may lead to new possibilities. Using a wire brush changes the way paint is thought of.

An artist or designer is not a craftsman for this very reason. Craftsmen are goal-oriented and constantly perfect and adjust their methods, while artists and designers need to challenge their own preconceptions again and again by working disruptively—and away from methodologies that maintain everything in its place.

I've alluded to three types of tools that play a role in three stages of a design process. Each type of tool addresses broad needs. At the outset of any design process, there is a need for tools to aid in casting a wide net, opening possibility and helping defer closure, and nourishing the sort of ambiguity that triggers the unanticipated. Thus, the first type of tool includes those for conceiving, experimenting, and generating (such as a brush). Second, we need tools for depicting (a pencil, for example). At the other end of the design tool spectrum, we need tools of production (among them the saw). Production tools are usually narrow and specific in function, far more resistant to unconventional uses, and are employed to achieve a specific outcome—not as part of a free-ranging process. These categories are not distinct: there is plenty of crossover, and a brush can be used for production just as a pencil can be used for nonrepresentational experimentation.

The relationship between tools and material or media is another part of the hand/mind/tool dynamic. Beginning a design process with a block of granite obliges a designer to use specific production tools and use them precisely. As with other types of fixity or affordances, materials and tools tend to be complementary, and so even beginning with a piece of paper will elicit its associated instruments. After all, why would you try to draw on paper with a chisel? However, tool and material pairings are flexible, particularly when the materials are easy to manipulate. Their malleability makes such weak materials sensitive and revelatory to the actions of the hand and to direct unmediated making, which can inspire impulsive "in the moment" acts.

Conversely, a strong material that strongly resists, that is unresponsive to action, can be drawn into a conversation only by deploying specific tools (which can play a catalytic role). To work with a block of granite requires training and experience, a period of embedding knowledge through practice, trial, and error. At the point of mastery—and here I would define mastery as the sublimation of effort in tool/hand coordination—the granite block yields to action and can act as an agent of creativity. The challenge with tools requiring such a high degree of skill and the time/effort to master them is that they will tend to have a conservative influence on the maker's work. In other words, the tools will tend to be used as they were intended—that is, conventionally—and less as a provocation of possibilities, of mistakes.

The phenomenon of being "shop-bound," caught in the fixity of tools and methods, is one of the consequences and challenges of a material-based creative culture. A loosening of such a fixity can happen only when designers give themselves over to the material at hand and let it—and not only the tools and methods—inform their thinking and their imagination.

DIGITAL TOOLS

The conserving, conventional coupling of "correct" tool with "correct" material affects the digital arena even more forcefully. Digital tools are ubiquitous in production, where a high degree of alignment between tools and operations is necessary. While machine tools tend to be material-specific (for instance, a Bridgeport lathe is only for metal), digital production tools are ostensibly more open to material variety because their interaction with material is fully controlled digitally—a laser cutter, for instance, can cut metal, wood, cardboard, or plastic. The fixity between material and machine seems to dissolve with the use of digital tools: water jet cutters, laser cutters, and multiaxis milling robots all accept a whole range of materials, including metals, plastics, wood, stone, and so on. Paradoxically, however, and unlike traditional machine tools that anticipate and are designed for specific material characteristics, these digital technologies accept a wide range of materials but are intolerant of any given material's behavior. The precise action of digital equipment on physical material depends on an idealized materiality, absolutely consistent and flawless. Anything these digital tools handle must be extremely uniform and unchanging—and hence the ideal digital material is, in fact, *immaterial.*

Design visualization software allows for generating and manipulating geometry and form expressed as line, surface, and volume. This has made it a powerful and relevant tool for the latter part of the design process, in which the project metrics are required in preparation for output and/or construction. However, as this software has shifted toward addressing the incipient phase of design, its effect on the design process has not been all positive. Digital visualization demands that even the designer's first act be dimensional, in effect refusing the immeasurable a place at the table. This naturally cleaves the design process: gestures versus keystrokes, intuitions versus commands, qualia versus quanta. This divide can be too wide to bridge, and so, increasingly, students and designers opt to enter the design process by embracing an almost purely digital approach.

At first, a digital approach can be reassuring, since decisions are made in simple terms: How long? How wide? How high? With simple and easy-to-use software, the results are not surprising; after all, the visual language of the software is inescapable and insistently generic. But why so, if outside of the software a mere 26 letters can create the immense riches of language?

Compared to the way letters combine to fashion words, joining visual elements has little or no significance unless there is a broader understanding of the larger whole. Putting a cone atop a cube doesn't inherently mean anything. For something to be meaningful, whether word or image, it must link to a concept. The design process is a search for meaningful ideas and concepts in the form of things and images, a search that mines the reciprocity of the two—that is, concepts arise from images as much as images arise from concepts. The difficulty in working with visualization software as a generator of ideas is that it lacks a means by which to link image to concept—something the thumbnail sketch of hand drawing elegantly achieves.

Using a tool is only a negative if we are convinced we can accomplish something that simply is not within that tool's capabilities. Awareness of every tool's incapacities empowers artists and designers to make conscious choices, including the misuse of tools as part of the creative process. Non-digital tools give feedback, generating a kind of tool awareness almost completely absent from digital tools, whose power and hidden apparatus make them immersive—a world unto itself, a world more suited to deterministic operations than to the back-and-forth search for concepts and relationships. Digital visualization functions a little like the apple drawing in chapter 4 in that it responds directly to a given command, the directness of which limits its ability to trigger richer wanderings of the imagination.

ACTIONS AND COMMANDS

The famed "Verb List" of American sculptor and video artist Richard Serra (b. 1938) is a handwritten list of words he called "actions to relate to oneself, material, place, and process," comprised of 84 actions and 24 conditions.[3] Of those, a handful overlap with the commands of digital visualization software: "to cut," "to fold," "to curve," "to rotate," "to twist." It is no coincidence that these few are all geometric operations. In comparison, most of his other actions, such as "to drop" or "to spill," have no geometric properties, are more disposed to the complexities and vicissitudes of material events and behavior, and are thus inaccessible to digital visualization.

Science has long since split along similar lines. Stephen Jay Gould referred to the two sides as, "for want of accepted names," the A and B sciences. The A sciences, such as chemistry, are "the traditional experimental, reductionist, nonhistorical, largely deductively oriented fields that try to achieve prediction based on simple chains of cause and effect (science with a big S of the classical stereotype."[4] In other words, they are the sciences whose subjects are reducible to time-reversible laws (such as Newtonian physics). Such laws are time-symmetrical as well; it doesn't matter whether time runs backward or forward. These sciences are highly reductive, building up and directing their attention to the interrogation of a hypothesis-based model of reality.

Merleau-Ponty, in his essay "Eye and Mind," referred to the A sciences—but might as well have been referring to digital visualization—when he wrote: "Science manipulates things and gives up living in them. It makes its own limited models of things; operating upon these indices or variables to effect whatever transformations are permitted by their definition, it comes face to face with the real world only at rare intervals."[5]

As to the B sciences—geology and paleontology, for example—which "do not traffic in prediction," Gould observed: "They represent a different and equally acceptable way of knowing appropriate for a large class of scientific phenomena. They require special intellectual skills (of synthesis and tangential thinking) that many people, scared stiff of science because they view it only in its A stereotype, possess in abundance."[6] The B sciences are informed by A sciences such as chemistry and physics but are not time-reversible; the history of the earth is a unique and nonrepeatable event.

Gould would often say in his lectures that were we to run the recording tape of history over from the start, the outcomes would be very different. B sciences must constantly mediate between causality that is event-based, such as dinosaur extinction, and causality as defined by scientific "laws," such as the formation of quartz. Meteorology is a similar example—weather obeys laws of science but is also heavily event-driven, defying prediction more than a few days out.

Thus science includes two antithetical goals: to understand causality from a cascading chain of unique, even indeterminate events in time, and also to understand causality as the regular and relentless playing out of the deterministic laws of nature. Material and geometry, Serra's list and digital tool commands, have an analogous relation: "to drop" is an event that resists determinism, whereas digital tools resist indeterminacy.

All design processes involve active mediation between the transformation of geometry to materiality (making) and of materiality to geometry (metrics). The rich difficulties with which geometry and materiality confront each another are central to design, and can be seen in the initial steps of any design process. Either geometry or material inform the work, but not both; even a gestural sketch is biased toward one or the other. This isn't only an issue of design process. It is one of history as well: the development of projection drawing in architecture uprooted the traditionally central role of construction (materiality) in the design process, a five-hundred-year trend that has accelerated with the advent of digital visualization tools. Today it is considered a serious flaw for a construction outcome to differ from instructions (drawings); the direction of influence has become exclusively a one-way arrow pointing from drawing to material, reinforcing the troubling preconception that our mind has a privileged agency to order the world without reflection.

Sigurd Lewerentz's decision to build his church in Klippan, Sweden (see chapter 7) without allowing a single brick to be cut was a decision to point the arrow of influence the other way, from construction to drawing. The architect was seeking to recover the lost reciprocity between material and mind, construction and design.

MOSAICS AND PIXELS

Digital visualization has come a long way since the days of glowing cathode ray tubes, Casio watches, Pong, and wire frames and pixelated images.

Amidst the universal admiration of the incessant advances of digital imaging, our changing relationship with images has gone unnoticed.

Originally, images were significant. As two-dimensional encodings of four-dimensional phenomena, they functioned as symbolic interpretations of the world "out there." The Czech-Brazilian philosopher Vilém Flusser (1920–1991) drew a distinction between such images and technical images—that is, photographs and now computer-generated imagery—and observed that whereas pretechnical images require mental scanning, reading, and decoding, technical images act more like windows into (or through) which we can simply "look." His view was that images had lost their role in culture as signs to be read, as abstractions of reality, becoming instead hallucinatory in that we restructure our reality into the technical images we are busy looking through, having forgotten that they are abstractions.[7]

The way in which an image is made has significance. Comparing a nontechnical image with a technical one reveals differences in interpretation, perception, and meaning:

I have long admired Roman mosaics as masterworks of design (figure 11.1). Why not call them "art"? After all, they are unique images made with a medium—little stones—and like all mediums, the little stones affect the image. Unlike other media, however, the little stones are not manipulated technically (such as when a watercolorist employs techniques to achieve a result) or systematically (as in brickwork)—rather, they are a design *because* they mediate between two contradictory organizational challenges, that of the image and that of the stone packing. People admire mosaics and many persist in calling them "art," but that is because they are so precious and are displayed in museum settings.

The best of the Roman mosaicists never applied a method of laying the stones; rather, they allowed the stones to be caught in the swirl between the autonomous image and the aggregation of pieces. Viewing mosaics separates these orders: the pure image is seen at distance and the stone order from very close up.

When mosaics are seen *in situ*, it becomes clear that they were intended as part of a larger design meant to be walked on, for instance,

11.1
Roman mosaic (left) and detail (right), late second to early first century BCE.

and not preciously held apart as "art." Apperceptions are most perturbed at the in-between scale where viewing of the image is not pure, where the mosaic is seen more as a design that takes into consideration the limits of the stone colors and sizes. There's a pleasure to be had in seeing the little stones still keep their identity and logic as arranged pieces while they transcendently suggest the glint of reflection in a glistening eye. In a similar fashion, faced with the limits of natural stone colors, the mosaicists—long before the impressionists—expanded their palette with adjacent colors of stones, which mixed together suggest—from a distance—a greater chromatic range.

Mosaics are linked conceptually to pixels by the idea that an image can be created from discrete and similar-size elements. Given the laboriousness of working with miniscule stones, there is reason for mosaicists to want to work with the largest pieces possible—while maintaining image fidelity. Comparing a pixelated image to a mosaic highlights the intelligence and economy of the mosaicist. With far fewer pieces, the mosaicist manages to create not only a smooth unpixelated image but a haptic one as well—not to mention the image's utility as, say, a finished floor. The significance of the mosaics is that they are signs of a way of living—the aspirations, concerns, subjugation, and skill of one mosaicist and of an entire society that has brought this work into being. That distinguishes the work from a pixelated image.

The pixelated image (figure 11.2) is pictorially simple, despite the technological sophistication of electronic computation. For the modern technical image to improve or advance, it does so with more computer power, more pixels, and more processing. The inertness of the pixel itself never changes; intelligence is located elsewhere, in the writing of the code and the engineering of the hardware. The stutter of the pixelated image annoys rather than intrigues—we are impatient to push past the surface to the image. There is no interest or tension between pixel and image. Paradoxically, despite the flattening effect of pixels, they do not pull the image back to the surface; rather, the image recedes as one is forced to back up in an effort to eliminate the pixelation and "resolve" the image. Only by vastly increasing the number of pixels will they disappear from sight and memory and thus allow the image to become truly hallucinatory.

What does an understanding of mosaics and pixels have to do with understanding how we think? I associate pixels with methodological

11.2

Pixelated photograph of former U.S. President Barack Obama.

thinking. They are reductive; they work through the application of rules. Mosaics are more process-driven and, like cooking, are materially engaged. The differences are similar to those between the reductive thinking of the engineer and the concrete science of so-called primitive cultures or the circumstance-bound thinking of the *bricoleur*. The differences are spatial as well: how we think on a screen is vastly different from how we think in space.

IDEAS AND KNOWLEDGE

Merleau-Ponty referred to "flesh of the world" to describe our being in the world as a fused and interdependent embodiment of self and all. In this world of the enactive mind, we no longer exist in the Cartesian *cogito* ("I think") separate from the body, but are whole as a correlate of the world. In this respect, the material imagination is made whole; the third orientation (which is not a dimension) is with the flesh of the world, into whose depth flows the felt totality of the image. The nature of our constructed reality, though, is such that we are pressed into the x, y, z of things, in particular the xy plane that confronts us, absorbing much of our attention—images, words, data, and measure all live in the xy plane of the world and increasingly dominate our lives and thought. Depth remains mysterious, subjective, the stitch that ties us to the world, resistant to *ratio*, to measure.

Taste, smell, and hearing direct themselves forward *toward* with little or no regard for the xy "plane of reasoning." Vision, left on its own, shifts its attention to the xy world, where the juxtaposition of things can allow for their manipulation and ordering. Only touch moves freely between the plane and the depth, limited only by reach. As Oliver Sacks (1933–2015), the eminent neurologist and author, wrote on childhood development, "This process of selection cannot arise, cannot even start, unless there is movement—it is movement that makes possible all perceptual categorization."[8]

Depth is the arena of movement, embodied perception, and enactment; it is there that space is fully synthesized out of all the senses. All of us know this from experience, as we move through space in an active bodily way. In sports, for instance, we sense the tug away from the inner-voice "I" and are enveloped in experience; attention is almost purely sensory and present-tense. How many times have we heard someone perform an athletic feat and remark that they "can't remember" what happened and how they did it? It is another instance of Heidegger's *Gelassenheit*, the process of losing willfulness, of letting go in order to be "caught in." It is important to note that the flesh is something we are *in* and that we can never observe it from a distance—it vanishes when we are looking at something as if it were a picture.

Technical images, particularly computer-generated ones, are *data*. As such, they are part of the seamless stream of information coursing across the *xy* of our screens. American journalist, historian, and film critic Neal Gabler has argued that thinking itself has been altered by our obsession with information, social media, web-based interaction, and internet "knowledge." Broad thinking, he claims, has diminished as we have moved collectively to information processing and a preference for "knowing."[9] Big ideas and conceptualization are less prevalent in this age of information because they require effort and time and an interest in the actual relationships between seemingly unrelated things (as in the mosaic example).

Conceptualization, the act of conceiving, is the moment/place where thought is embodied. Unlike data or knowledge, it depends on contradictions, anomalies, discrepancies, and incongruities. The smoothness of the digital attractively unifies, but it also dulls the impulse to conceptualization.

Information processing has affected design thinking at the point of inception. It has become fashionable to overlay many discrete diagrams of information to generate an automatic or found order. Each diagram constitutes a single system, and the juxtaposition of the diagrams suggests an order more authentic and complex—a whole constituted of many parts.

The similarity of this kind of diagramming with design software (layer management) is hardly accidental: layers are the architecture of visualizing software. But juxtaposition is not organization. Life does not parse itself into layers of order that seamlessly overlap. Two kinds of logic—let's say a logic of structure determined by an understanding of loads and forces and a spatial logic that is informed by use needs, movement, and the choreography of experience—may seem like good candidates to superimpose. As every architect knows, however, these two contradict each other and *because* of this have driven architectural thinking for centuries. Such *irresolution* keeps architecture and design relevant and an enduring proposition. As with conceptualization, only when disparate and contradictory parts are brought into a meaningful relationship can there be organization.

THE DIGITAL HAND

In "Representation and the War for Reality," William H. Gass wrote, "The problem was, in effect, to unite the two: to introduce mathematics into the confusions of observation, and the loud rich tumult of the world into the thoughtful reticence of angle, plane and cube. Apart, one was blind, the other empty."[10] This describes well a task I undertook to bring measure to the phenomenal world and drag observation into the concepts that can live in the computer.

A few years ago, I worked with three of my colleagues—Kyna Leski, Carl Lostritto, and Pari Riahi—to redesign our department's drawing curriculum, which had become dated and bifurcated into a hand drawing course followed by a course in digital drafting. As might be expected, students' hand drafting had fallen into disuse as they became increasingly adept with digital visualization tools. We asked ourselves whether there was still a role for the hand and, if there was, whether it and the digital realm could find a symbiosis.

Computation was the key principle for the new course. Computation—regardless of tool, be it pencil, computer, or brush—took on the role of language. It became a translator between the qualitative and the quantitative. Students were asked to compose a limited still life and draw it from two distinct points of view. These drawings were then projected back onto the still life, opening a new visual/phenomenal field operating *between* the still life and its representation (figure 11.3).

The students were challenged to transfer all the phenomenal aspects—color, shade, texture, reflections, and so on—into the computer, the half image/half still life serving as a critical hinge between the qualitative and

11.3
Still life aligned with projected image (left) and oblique view (right), Tida Osotsapa.

the quantifiable aspects. To the careful observer, there is no end to what can be seen. How, then, could each student bring measure to his or her rich work without getting buried in the sea of data points or reducing it to caricature?

Lou Andreas-Salomé (1861–1937), a Russian-born psychoanalyst and author who counted Nietzsche, Sigmund Freud, and Rainer Maria Rilke among her friends, wrote in her memoirs:

> Whoever reaches into a rosebush may seize a handful of flowers; but no matter how many one holds, it's only a small portion of the whole. Nevertheless, a handful is enough to experience the nature of the flowers. Only if we refuse to reach into the bush, because we can't possibly seize all the flowers at once, or if we spread out our handful of roses as if it were the whole of the bush itself— only then does it bloom apart from us, unknown to us, and we are left alone.[11]

The mosaicists, extending the limits of their means, offered a lesson in how. By mapping graphic orders onto the still life, not unlike the way Roman mosaicists ordered their little stones like a kind of semiautonomous tattooing, the students got to their Andreas-Salomé "handful." Aided by new tools and templates and gauges, the students transferred the ordered measurements into their computers and slowly constructed a digital model.

The digital architecture of the software invited measurements to coagulate into figures, patterns, lines, and volumes, each separated onto a discrete digital layer. A transparent glass (figure 11.4), for example, could be defined as a set of mapped reflective figures or as a careful point cloud

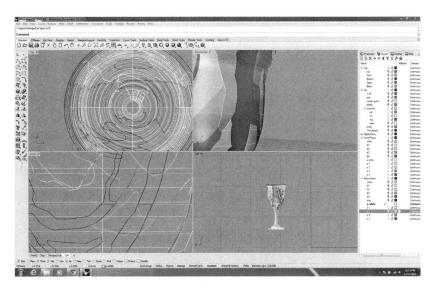

11.4
Screenshot of transparent glass, Jenyea Chang.

of measurements (which doesn't resolve into anything other than a cloud unless connections, such as lines, are made between points).

An interesting thing happened once the cloud of data resolved into even the slightest degree of recognition: the students could seemingly reach into the data and begin to see and elicit the still life. The glass could also be reconstructed digitally as a set of geometric shortcuts; circles, cylinders, cones, and so on could be quickly marshaled to synthesize the glass with, at most, a cursory glance at the still life. Percepts were trumped by concepts. The digital layers were empowering, allowing all the logics to reside within the model without occluding each other. Some projects had close to a hundred layers, most if not all of them comprising overlapping logics. The final images (figure 11.5) were extractions from the model, a mining of layers and a unifying action through which the phenomenal aspects of the still life and new unanticipated consequences of the digital process were commingled (figure 11.6) and synthesized into a new hybrid vision.

Chance and naiveté are sources of creativity—unearned perhaps, but nevertheless important ingredients of the digital work. That and the messy

11.5
Screen shots of digital model (left) and final extraction (right), Tyler Mills.

11.6
Digital model, Yitan Sun.

mixing of concepts, tools, point measurements, and templating provided sufficient disruption to trigger imaginative visual speculation. In the best work, students felt a keen sense of authorship, that they weren't imitating the visible. It was an affirmation of Paul Klee: "Art does not reproduce the visible; rather, it makes visible."[12] The computer proved powerful in forcing consciousness, in particular consciousness of line, to the foreground. Without lines, the point cloud—a set of data points in space, a phenomenon particular to the computer—remained amorphous, leading the students to discover order by creating lines between points, lines that "rendered visible."

Shelley's definition of the imagination as synthesis follows the words of German philosopher Friedrich Schelling (1775–1854) in his lectures in Jena and Würzburg at the dawn of the nineteenth century: "The splendid German word 'imagination' [*Einbildungskraft*] actually means the power of *mutual informing into unity* [*Ineinsbildung*] upon which all creation is really based. It is the power whereby something ideal is simultaneously something real."[13]

Schelling points to the real value of the students' work: that it provided enough tension between the ideal—the computer's reductive geometric, even eidetic tendencies—and the real "stuff" of the still life, with all the messiness of the measure of those *qualia*, to bring forth the unifying force of the imagination. Bachelard might have called it a dance of the formal and material imagination.

12 SMOOTHNESS

Who hasn't enjoyed that moment when the long wait in line melts away or something is accomplished with the mere press of a button that only a year earlier took much more effort?

While labor-saving devices are as old as the stone axe, the dream of reducing daily physical effort with helpful devices dramatically accelerated with the advent of electrification. Today, living in the midst of the digital storm makes it hard to know where we are headed, but there is a direction. The dream of saving labor continues but is now marked by a desire for smoothness: a smoothness of experience, of effort, of movement, of adjustment, of communication, and the connected smoothness of data flowing to, through, and from. The screen and its images are gaining ocular certainty and a smooth connectedness to experience. If it's on screen, it's "real"—a reality that in the past was filtered by a professional class in academic and cultural institutions, broadcast, or publishing interests, but today can be "curated" by anyone.

Given the ubiquity of seamlessly flowing virtual imagery, text, and data, and the brain's finite capacity for mental processing, smoothness is not only a dream but a sensible strategy for reducing cognitive loading. Fixity, preconceiving, passive engagement (not using the body), and stable *Umwelts* that support functional extensions of mental processing (such as the airplane cockpit of chapter 9) join with smoothness in absorbing the new demands on our attention engendered by our digitally networked existence, which competes with physical experience as a primary coupling of mind. The challenge of the new smooth model of existence is to leverage mental capacity using digital resources while not diminishing the critical role of material engagement, which couples us to our physical *Umwelt*, nourishing the imaginative and creative mind.

How is resistance to be squared with smoothness? After all, humankind's progress came from efforts to make, to articulate material, coupled to that material's resistance, which has continuously brought forth our creative and imaginative potential, providing a fundamental grounding and meaning to life. To understand what has happened to making, it is worth clarifying how the way we make has long defined the role of material, and how traditions of making by hand once depended on a symbiosis between maker and material—a symbiosis that disappeared with the introduction of mechanization.

FROM MAKING TO CONSUMING

The relation of making to humankind, rooted in the relation between material and making, matter and form, has been a philosophical subject for millennia. Aristotle, in his book *Physics*, defined as the matter of *x* whatever *x* is made of (which in the nineteenth century came to be called *hylemorphism*, a word created from ὕλη—*hyle*, meaning "matter," but also used for "wood"—and μορφή—*morphē*, meaning "shape," "form," or "appearance") as he sought to establish what being is.[1] Aristotle arrived at the position that matter was mutable and form was the unchanging, actualizing principle. Form in this sense is not reducible to shape but rather is the result of something essential being materialized.

Aristotle discussed causality in his *Metaphysics*, and identified as one of the causes the relation between form and matter in the act of making.[2] Two millennia later, Martin Heidegger described the fourfold causality underlying an object, using a silver chalice as an example.

> (1) the *causa materialis*, the material, the matter out of which, for example, a silver chalice is made; (2) the *causa formalis*, the form, the shape into which the material enters; (3) the *causa finalis*, the end, for example, the sacrificial

rite in relation to which the required chalice is determined as to its form and matter; (4) the *causa efficiens*, which brings about the effect that is the finished, actual chalice, in this instance, the silversmith.[3]

These four causes are not independent but stem from one another: the material cause, the silver, informs the shape of the silver chalice (*causa formalis*), and its final use also informs the shape. The *causa efficiens*, or activating agent, is the maker, who does not impose on but receives the other causes as part of the maker's agency. As long as making is by hand, by crafting, the maker meditates with material, bringing forth its form, its "truth" or unconcealment.

Walter Benjamin used the term "aura" to describe a work's metaphysical authenticity arising from this intimate dialogue between maker and material.[4] Once machines replaced the hand, form no longer emerged from matter; rather, it was *imposed on* matter. This changed the tacit understanding of material and its previous participation in the actualization of form, as well as inverting the craftsman/artist's meditations from and through material since, with machines, abstract thinking precedes the process of making.[5]

The power of mechanized making turned attention from the meditations of the craftsman/artist to the forces and pressures and mechanical motions of machinery, controlled by new languages of calculation: engineering, mathematics, and physics. As the hand disappeared from making, the accompanying demise of "aura" was compensated for by the power of mass production and reproduction.[6] Form and matter could no longer be reduced to a simple binary; rather, as French philosopher Gilbert Simondon (1924–1989) outlined in his seminal *On the Mode of Existence of Technical Objects*,[7] form was distributed throughout the technical ensemble needed to make something. Even a simple sewing needle required a mold, a stamping, a polishing, all agents of the technical assemblage, which encompasses not only the needle but the sewing machine, the textile machinery, and so on. Material in such an assemblage of processes, which define mechanized production, is always in play, a participant in the making, not singular in nature but varied and interdependent with other materials: a mold must be of a harder material than the object being molded.

Throughout the development of mechanized making, science and technology have steadily redefined material toward something understood, controlled, and made wholly predictable, including synthesizing it and creating it from irreducible and artificial stuff. This affects our attitude toward materials, shifting us from a more empathetic and reciprocating relationship (making by hand) to one characterized by intentionality and the imposition of uniformity and control (making by machine).

How, then, does this description of making intersect with design, since designers are neither artisans nor engineers? Design spans between craft and machine. The designer makes and mediates with material, acting as an artist/craftsman, but the making in this situation is a model, or maquette, or representation of a future machine-made form (form as an actualized essence), whether a shirt, or tableware, or appliance, or building. The clay modelers carving automobile models are one example (see chapter 1); this corresponds to the meditative exchange between craftsman and material. Though the cars are subsequently mass-produced, there is enough belief in material-based design's capacity to transfer some "aura" into the machine-made result that automakers continue to rely upon handmade prototypes.

For millennia, the challenge of material's resistance was centered on the body. Countless souls and civilizations have undertaken the hard, physical work of articulating the world. In the last centuries, technology—the means a society uses to provide for its material exigencies—has gradually displaced the body from its central and direct physical role in making or production. And seemingly with each passing day, production moves further afield and becomes further disengaged from daily life.

Highly manufactured objects no longer reveal their process of making or logic of assembly—often expressed by their irreparability. This has been accompanied by a corresponding shift from cultures of making to cultures of consumption, which has affected our coupling to material. Consumption of material goods does not generate resistance; that has to be found elsewhere, such as in sports and recreation, which, as the latter term implies, are means by which we *re*-create the lost natural physicality of our species. Fortunately, hobbies, handiwork, and other avocations such as cooking and gardening continue to thrive as meaningful engagements with the resistance of material, innately understood. The fading culture of physically working with material doesn't mean resistance has disappeared altogether.

It also doesn't mean Earth is irrelevant or uncoupled from the World's articulation. The Heideggerian strife between World and Earth continues; however, the resistance of Earth now operates at a distance from the objects that articulate World. The danger in this modern form of unconcealment—the *aletheia* of our times, with its pollution, shrinking species diversity, deforestation, water shortages, climate change, and resource depletion—is that it all continues apace while the objects articulating our World are no longer perceived to be related to, a cause of, or at fault.

SITUATED AND METHODICAL THINKING

Mathematicians, computer scientists, engineers, and software developers work with challenges and difficulties in their respective fields. Just as working with material does, these (non-material-based) challenges affect their ways of thinking.

From the viewpoint of the extended/enacted mind, both our articulation of our world and the shift away from material-centered work have had an effect on mental processes and thought, away from what I have named "situated thinking" and toward something more linear and planned. I call this "methodical thinking," which can be characterized as model-based or tool-based, purposeful, and involving methods of operation or logic. Comparing the two should ring familiar, since both ways of thinking are part of our mental and physical landscape and a result of our evolutionary adaptation to the environment. Here is a quick comparison of their features:

Situated Thinking	Methodical Thinking
Enactive mind	Extended mind
Material/*Umwelt*-based	Tool-based, applied method
Provisional/contingent/situational	Oriented to purpose
Rule-breaking	Framed/planned/defined/methodical
Qualitative	Quantitative
Narrative	Reductive

Situated Thinking	Methodical Thinking
Circular / recursive	Linear
Imaginative (synthesizing)	Analytical (breaking down)
Self-knowledge/authorship	Fact-based knowledge
Tacit understanding	Technical understanding

This isn't an either-or proposition; everyone thinks in multiple ways. A software developer is capable of cooking a chicken *and* scripting a long series of logical steps of code. Experienced brick masons use plumb lines, rulers, trowels and pointing tools to maintain their aim of consistency and evenness in laying a row of bricks (methodical thinking), while continually adjusting (situated thinking) the amount of mortar placed on each brick gauged by the mortar's stiffness, the brick's dryness, air temperature, humidity, and the time of day to achieve their craft.

The *way* of thinking changes in every circumstance. As we become less and less connected to material, methodical thinking gains prominence. In a world more attuned to data manipulation than material manipulation, situated thinking is being given less credence. In the process, the participatory role of material and the *Umwelt* in thinking, imagination, and empathy is lessened, and the understanding of what constitutes thinking narrows.

RESISTANCE AND MEANING

Digital fabrication has upended traditional ways of making and in the process has revealed how much meaning arises from resistance. Previously, the meaning of making came from the difficulty of working with a complex situation, whether it was an Inka carving a perfectly fitting polygonal stone or a craftsman making a silver chalice or even the many interdependent steps and machines needed to make a designed object. In architecture and design pedagogy, there is a tradition of introducing resistance, creating difficulty or limits in order to elicit a meaningful outcome. One classic problem was to challenge students to design a chair, limiting the available material to a single sheet of plywood. Such pedagogy, though not a challenge likely to be encountered in the "real" world, was seen as instilling the values of efficiency and inventive problem solving into thinking about design and, in so doing, making the work meaningful.

The design by Danish architect Jørn Utzon (1918–2008) of one of the most celebrated buildings of the twentieth century, the Sydney Opera House in Australia, is a testament to the symbiosis of meaning and resistance in architecture. When Utzon won the competition in 1957, his proposal was barely sketched out, without much more than an ingenious planning solution to a difficult site and the now iconic image of the Opera House roofs expressed as free-form, sail-like shells (figure 12.1). In order to build this monumental structure, Utzon had to translate the gestural sails/ shells into a structural logic, without losing the poetic lyricism of the forms.

Utzon spent years applying architectural solutions to the site problem, but he and his engineer, Ove Arup, found the resolution of the formal, structural, and constructional geometries of the shell designs were not converging on a single design. The problem they faced was intensely resistant to a solution—until, in what is now memorialized in countless retellings, Utzon had an "aha" moment in his studio, finding a simple geometry that could resolve the constructional problem in a poetic form, and that also resolved structural difficulties. He grasped that if the surface geometry of the Opera House shells could be made uniform and unchanging, it would

12.1
Sketch of Sydney Opera House prior to spherical geometry solution, Jørn Utzon, 1958.

be far simpler to build—like a module being repeated over and over. But there was a further problem: with such a uniform geometry, the imagery of unique and separate "sails" would suffer.

Utzon's solution was to propose that all shells be geometrically developed from a single sphere with a radius of about 247 feet, but that each shell segment be extracted in such a way as to fit the varied sizes needed for each shell and still appear independent of the others (figure 12.2).

Once the spherical geometry was adopted, the engineering became exponentially simpler, allowing for a precast process through which shells were developed from a series of precast arced ribs like the segments of a collapsible hand-held fan. (Whereas standard concrete is poured into site-specific forms and then cured on site, *precast concrete* is produced by casting concrete in a reusable mold or "form" and then curing it in a controlled environment, after which it is brought to the construction site and put in place.) Though each rib for each shell was unique, all the ribs could be made from the same mold (a portion of the 247-foot radius), a highly efficient fabrication process. The final difficulty—how to cover the shells with ceramic tiles—was solved once the surface geometry was simplified to that of the sphere.

The design solution for the Sydney Opera House roofs is still considered one of the great design moments in modern history. The architect, working with the engineer, slowly puzzled through the enormous

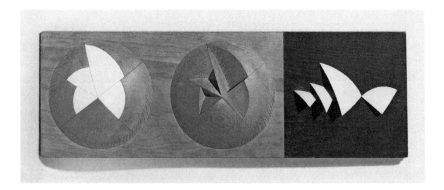

12.2
Utzon's shell geometry extracted from a sphere, 1962.

resistance of the problem and forged new architectural relations between contradictory conditions, between gesture and geometry, between modularity and movement, between simple elements and complex form. The shells of the Opera House are a triumph of design because, like the chair made from a single sheet of plywood, they resolve the intractable resistance and complexities of the problem with an irreducibly simple solution that integrates all aspects of the shells. It is impossible to regard any component of the design independently from the overall construction. The integrity of the whole, the architecture, stems from this totalization of form/geometry/ structure/construction/image. The meaningfulness of the building emerges through the very act of overcoming resistances and making complexity simple. This totality has been celebrated as such and has extended the meaning of the building, its greater sense of connection and relevance, to symbolize a city, a country, and a people.

It is worth noting that while the process of solving the challenge of the shell problem involved many strategies and failed attempts, it was situated thinking—thinking that came *from* the material and models at hand—that led to the breakthrough. As Utzon later related, in order to free up some space in his design studio, he gathered together assorted earlier paper and cardboard shell models, each made from a unique set of curves. When he saw them stacked together, they looked remarkably similar, leading him to the insight of deriving the shells from a shared geometry. Had the situation not presented itself this way, had the shell models remained separate, preserving the fixed conception of their being unique and individual, had Utzon not accidentally piled them up as a set, had there not been so many failed models in the first place to take up all that space, the breakthrough could not have happened at that moment.

DIGITAL FABRICATION

The Sydney Opera House shell construction process involved one of the first uses of computers in building design. Complex calculations were required to ensure that each unique rib, as it was being erected into place, would not carry or transmit excessive forces. In those days (early 1960s), computers had no bearing on design outcomes, but today digital technology dominates the design and fabrication process. Highly complex surfaces and elements can now be modeled and immediately sent to be digitally fabricated, bypassing human hands. The manufacturing industry relies on rapid prototyping to test-fit parts for complex assemblies almost instantly, and the dream of digitally "printing" metal ware, buildings, and even complete mechanisms with moving parts and multiple artificial and intelligent materials have come true.

In short, it has become easier to make complex things, smoothing away many of the former difficulties.

If the Sydney Opera House were designed and built today, there is a vanishingly small chance it would be constructed with the same simplified geometry. Why is that?

When Utzon began working on the shells, he bent a plastic ruler as a way to start a controlled process of making curves for the ribs. He sent off versions to the engineers, based on formal and architectural concerns, asking them to resolve the structural issues. It took years of work that slowly brought the multiple dimensions of the challenge—the structural, the architectural, the constructional—into close enough correspondence for Utzon to have his epiphany. Had he worked with current digital tools, the resistance of the problem would have receded immediately.

Imagine the architect digitally sketching a roof shell to fit an architectural intention, using complex vector manipulations to achieve a flowing digital result that circumvents traditional *xyz* spatial conventions, and then sending it off to engineers, who muster their sophisticated engineering resources to analyze the complex forces, optimize efficiencies, and solve the structural issues. The construction itself would also be less resistant, since with the advent of digitally controlled fabrication processes, building each shell out of thousands of unique, one-of-a-kind parts is no longer the problem it once was. The resulting constructed shell would likely be far more complex in form and parts. But would it have been a better work of architecture? Would it have been meaningful? Would it have become a symbol for the nation?

This is the situation today. The interdependent relationships among material, form, structure, construction—inherent in the natural world—are no longer the source of meaning they once were. Design is increasingly propelled and shaped by technological means, rather than by the mysteries of the human condition. The meaning of the Sydney Opera House is rooted in its essential humanity: the creative intelligence, sensitivity, and imagination of its authors, all of which, it should be added, was brought forth by a material resistance to the resolution of the design problem.

Tools have a way of pulling us toward their capacities. Digital fabrication tools are no different: vanguard experiments in digital fabrication often occupy territory marked by a proclivity for the biomorphic, complex topological surface geometries, and atectonic assemblages (in which the individual parts are subsumed to a larger order, giving the appearance of a monolithic construction) made from (almost) seamlessly joined unique individual parts that often incorporate cascades, waves, ripples, wrinkles, twists, alluding to a reactive, vital responsiveness, as in the example in figure 12.3. Such complexity, heretofore almost impossible to achieve, seems justified if for no other reason than that these tools make it possible. Like cinema in its infancy, the technology dazzles enough that content is not at issue.

A curious characteristic of these digital experiments is that they no longer seem authored or the result of a process of thinking and making. They gain their autonomy as things in the world, without any particular communion with humankind. Digital fabrication expresses an aspiration to a timeless and dematerialized flawlessness, and the consistency of this vision worldwide convinces me that digital procedures, tools, and software have a powerful influence on the design intentions of authors who employ them in their work. The influence, though, is not the same as that of material that has an apophenic potential—the human tendency to "unmotivated seeing of connections [accompanied by] a specific feeling of abnormal meaningfulness."[8] I am making here a distinction between material's role in catalyzing imagination, expanding creative potential, and tools' affordances and role in guiding and narrowing intentions: "If all you have is a hammer, everything looks like a nail."[9]

12.3
La Voûte de LeFevre installation, Matter Design, digital fabrication, plywood, 2012.

Until digital fabrication, making had been understood to be subject to material properties; even the crudest assemblies incorporated tolerances and specific details to intelligently anticipated material behaviors. And even (as in the case of architects) where the initial model was immaterial, as with blueprints, makers and builders translating the drawings brought materiality into the fabrication process through their experience, tacit knowledge, and craft. Digital models, though, forget material and tend to idealize the relation between parts. An impossibly perfect fit and digital fabrication tools propagate this inherent bias toward perfection: the cutting heads, water jet, laser, spray nozzle all want one thing, which is to have material behave digitally, reducible to data.

Ironically, digital fabrication manifests the flawlessness of complex digital formal models (figure 12.4) as material flaws, often encountered as small but annoying misalignments. No matter how precisely parts are shaped as part of an assembly, they never fit exactly as in the digital model, since the model makes no allowance for the inevitable warping, shrinking, spring-back, compression, and expansion of material. When natural material is subjected to the digital process, its vitality remains undiminished, an echo of Heidegger's resistant Earth.

12.4
Digitally fabricated and joined screen elements, Maison Folie Wazemmes, Lilles, France, 2004.

The tools of the digital fabrication process are among the least responsive to material and intolerant of any inconsistent reaction, effectively negating material. Can the digital process become a reciprocal process, one in which material plays an integral part?

There are ways. The first would be to create a uniform material closer to the digital ideal in order to align the digital model with materialized fabrication. In such a closed system, there would be no need for adjustment or translation. A second way is to bring design and material knowledge into digital models—similar to the way assemblages, from automobiles to buildings, have incorporated details and tolerances in anticipation of material behavior and transformation. Another way, now standard practice for engineers, is to model material properties in software; however, these properties are limited to quantifiable force behaviors such as bending, buckling, and shear (in the case of structures). For now, the qualia of material remain outside the digital realm, but interesting materials have appeared which can be digitally printed and then shaped by hand, opening the possibility of a design object being shuttled back and forth between a digital and physical state, emerging as a hybrid of material and data.

Yet another way for material to enter the digital system is to develop output tools that are more responsive to and tolerant of the uneven qualities of natural material. Imagine a digital tool "feeling" the grain of the wood being shaped and adjusting the tool path in anticipation of the warpage and shrinkage that will inevitably follow. That is the territory of craft.

During the Renaissance, when projection drawing entered the practice of architecture, it disrupted the medieval locus of building design, which was of and in situated action, guided by the tacit knowledge of craftsmen and the onsite choreographing of work and material, replacing it with the idea of drawing the building as the arena of design. Digital modeling software is now challenging the concept of the static drawing, with new tools able to stretch, wrap, move, distort, and twist surfaces into complex topologies. Performative-based design, a digital-based concept, envisions the transformation of the static built environment into one that responds to and "performs" with and for its inhabitants. The concept is not new—it's actually as old as the oldest building in that humankind has forever been adding performative design to the environment. A window moves and adjusts, as does a window shade, as does a heating and cooling system. What is new is the idea that data-driven digital modeling and fabricating will impart an almost infinite flexibility to the otherwise static built world, a complex flexibility that approaches that seen in life, such as with shape-changing skins and heliotropic elements. The dream is to model and articulate the world as a continuum. Terms such as "touch of a button," "fluid movement," "smart materials," and "automatic responsiveness" allude to a smooth, effortless, and sensuous encounter with one's surroundings, which are infinitely adjusting to optimal conditions of light, energy use, comfort, and so on.

Though some structures have incorporated technology that can shape-shift an entire wall, most of the built environment remains static. Digital modeling software may have generated the aspirations of performative design, because it can model such a world relatively easily, but there remains a persistent gap between the modeling and the technical manufacture of such a fluidly moving, adjustable, performative work.

The challenges of performative-based design and highly volatile, complex, smooth digitally described surfaces and forms lie in their connection to material, or lack thereof. When material doesn't participate in the genesis or propagation of new digital landscapes, it in effect limits the digital models to pictures. Removing materiality from digital software leaves a virtual world, one both pliant and pastelike, both easy and difficult to work in. It is easy because everything is manipulable, and difficult because greater ease makes it harder to identify problems or aims or possibilities when modeling, since meaningful connections require a certain level of recalcitrance and contradiction. The physical maker, who constantly adjusts in response to changing circumstances, is thinking in a situated manner; the digital modeler less so.

FLUX

Earlier in the chapter I posed the question of how resistance is to be squared with smoothness. In the discussion of digital fabrication and modeling, smoothness emerges as a promise, a vision of a more attuned and responsive built environment. Smoothness also emerges as a challenge to the imagination and our deeply felt need for meaningful relations between ourselves and the world we have come to articulate. Smoothness desires to remove the lumpy, the inconsistent, the awkward, everything giving rise to doubt, yet it is exactly those incongruities of our surroundings that call forth articulations and narratives to restore and complete our sense of being in the world, with the aim to make experience meaningful and plausible, not necessarily to concord with facts.

The richness of meaning that comes from making, from images, and from language is propelled by the flux of the *Umwelt* and by the means themselves. Making inspires images; images inspire words. Language is constructed from the fluctuating, volatile relationship between the physical world and mind; metaphors conflate material properties with mental concepts (as "I am reflecting on this"). If the arrow of a metaphor points from the concrete to the mind, pointing in the opposite direction are projection, conceptualization, and anthropomorphism.

Consider these characteristics of human personality: compliant, pliant, tarnished, stubborn, fake, authentic, phony, tough, strong, sharp, superficial, humble, firm, honest, crude, resilient. We readily project them onto material. A chemist would never agree that from a *scientific* point of view vinyl plastic is fake and brick is honest, but our culture says otherwise.

Conflating personality characteristics with material properties doesn't clarify which way the arrow is pointed, but it does suggest the influence goes in both directions. We find things meaningful because the social and cultural matrix is fashioned from encounters with the unfathomable nature of the physical world, encounters that extend and enact the mind and fuse culture and place. In receiving from and projecting onto the world they inhabited, the Inka formed a culture and social order that could not have existed elsewhere.

Flux is also in the mind as it extends and enacts and shifts between situated and methodical thinking. Flux between modes of thinking is not only valuable but even considered an essential part of military, aviation, and police training for mastery of sudden and unpredictable events, to which those stressful fields of operation are highly susceptible. Flux is like electricity, powering cognition, experience, and enactment of the self-world. The greater the disjunction, resistance, and anomaly in the self-world, the greater the potential to bridge with action, imagining, narrative and meaning, creativity, empathy, thinking and making.

In our quest for smoothness, are we removing the sources necessary to fully enact our self-world?

IMAGE, SURFACE, AND SCREEN

Until now, because of their inherent flux, images have been a powerful source of mediation between self and world. The first drawn image, acting both as a surface and as a thing seen through the surface, required engagement and personal interpretation. That is still true today, but each time pixel counts jump and images become more hyperreal, flux is reduced, rendering the surface mute. When seen on a screen, images, especially high-resolution images, tend to be more explicit, stable, fixed, and less open to interpretation. If every eye inherently sees differently, guided by individual experience and sensibility, then as we gaze collectively into the screen is the new fixity of images having us forsake the uniqueness of interpretive seeing for something more common, shared and monolithic, smooth and uncritically consumed? Can digital images recover the flux that animates the dissonances of surface and image? To address this question, let's consider images found in nature and images as authored in a medium.

The flux in and of images, besides that presented by surface, depends on content. A mark done just so might suggest a foot, or perhaps it suggests a tree root. Such ambiguity has always been part of the importance of images for the imagination. In addition, the spatial ambiguity of images, the flux between flatness and depth, is a phenomenon of painting and drawing, something that links to our perception of the world as we find it.

Pareidolia, the inclination to see images in ambiguous patterns of clouds or on the moon, is a trait of perception often characterized as a desire to find meaning. Such visual filling-in is probably more "hard-wired" and less about the imagination. However, pareidolia offers a valuable clue to an image's broader trigger for the imagination. Whether it is animals or faces or other familiar objects we recognize in a cloud, the common attribute is a flattening of depth. Clouds are common subjects not only because they are evocative but because they are far enough away to be viewed as image-like. Even as iconic a natural formation as the (now-collapsed) Old Man of the Mountain in Franconia, New Hampshire (figure 12.5) was always "seen" at a distance that rendered it more or less two-dimensional.

The suppression of the third dimension is essential to opening us to the powerful place images occupy between raw perception and thought, and as a product of both. The two-dimensional aspect of images frees them to act in multiple ways. An image can be understood as having depth,

12.5
Old Man of the Mountain, Franconia, New Hampshire.

invoked by clues such as pictorial foreshortening and diminishment, or conversely as flat. Signs and symbols, which hover between being images and writing, tend to appear in shallow images, which in turn tend to further flatten the field they occupy, such as this page.

Paradoxically, a flat image can bring attention to the materiality of its surface; in the case of painting, the texture, brushstroke, and building up of a medium are inseparable parts of the subject matter. Some painters have made surface *the* subject matter of their work. This ambivalence about materiality and the relation of self to world suggests to me that the "reality" of images is in constant fluctuation between being experienced directly as a phenomenon through all the senses and being "read" or interpreted as idea or representation. What images tell us is that materiality is both symbolic *and* phenomenal, suggesting a degree of integration of material with the mind that goes largely unnoticed.

Such a state of flux can be seen in a drawing by the French artist Georges Seurat (1859–1891) of the artist's mother (figure 12.6). Attention to the drawing as a "picture" is contrasted by the rich materiality of the Conté crayon (carbon black in a clay base) and paper. The paper was handmade in a paper mold, which yielded sheets with a pattern of parallel ridges; when rubbed with a medium such as Conté crayon, the result is a strong geometric pattern in which the crayon is deposited on the ridges, leaving the valleys light in tone. The image itself is made without any line or distinct marking, appearing as a pure, soft, atmospheric modeling of mass, light, and shade, an emissive insubstantiality growing out of the dense matrix of rubbed crayon (figure 12.7). The effect is absorptive, free from reflection, visually anechoic, a dark solid gently sublimating into light. A sense of calmness, of quietude, heightens awareness of the drawing as an optical phenomenon; this quality, independent of subject matter, can be sensed in all of Seurat's drawings done in the same manner. The impressionist painters of that period had sought to bring immediacy to painting, to capture the impression of the moment; in contrast, Seurat's work distends time, slows it, calmly holding our attention in a state closer to reverie or daydreaming. Why is this?

The contrast between the materialized texture of regular gridlines of the conté on paper and the ethereal tonality of the image creates something of an apparition—not quite there, not quite in the here and now. Move closer to the sheet and the haptic qualities dominate, the image vanishes, and the hand wants to feel the sensuous surface. Conversely, when viewed

12.6
Madame Seurat, the Artist's Mother, Georges Seurat, c. 1882.

from afar, the image is predominant and the perceptual perturbation, caused by the flux between patterned surface and luminous representation, abates. When seen in person, the drawing invites the viewer to move about, to experience what I would call a coming in and out of consciousness, the *process* of the various perceptions. This, I believe, invokes reverie.

12.7
Detail from *Madame Seurat, the Artist's Mother*, Georges Seurat, c. 1882.

When the multiple perceptions overlap, a new sort of depth appears: the image appears to recede behind the light horizontal laid lines of the paper, but only when attention is focused on the surface. The receding of the image is, in part, apparent because it is a representation of the familiar form of a woman; it also recedes from light, which (seemingly) irradiates the page from near our point of view into the dark depths beyond the surface.

The philosopher Gaston Bachelard wrote of material imagination, and Seurat's drawing operates as such. We are invited to probe mentally and dream into the drawing as a material phenomenon, one from which Seurat's mother appears as an immaterial apparition.

Seurat was an extraordinary draftsman, his early work marked by a hyperrealistic mastery of the figure and of drawing technique (figure 12.8). The relationship between his early drawings and their medium/material was familiar in the sense that the drawing dominates the medium—the paper was chosen for its subservience to the drawing process, with a fine grain that would not impose structure on the drawing surface or intrude on the foregrounded image.

As we have seen, when Seurat began to work with a handmade paper distinguished by its laid lines and chains (linear troughs running perpendicular to the laid lines), all that changed. The partnership of author and material propelled the imagination *because* the material participated in the making of the work.

Seurat is famous for developing pointillist painting, the building up of a work from points of color. His drawings were the source of this, the texture of the paper breaking the Conté marks into a textured field of points. Would he have been able to find his pointillist approach to painting without the effect created by the material of the paper and conté? There is no evidence to suggest otherwise, but it also bears noting that Seurat's extraordinary ability to draw heightened the contrast between his exquisite renderings and the assertive paper-induced texture, enough to trigger the conceptual breakthrough of pointillism as an autonomous logic. Pointillism as a phenomenon preceded pointillism as a concept; material behavior led to new thinking.

Though Seurat's pointillism presaged digital images, his work engages vision differently. Whereas his drawings bring forth apparitions, digital images do not; they are hallucinatory in the sense that we view a digital photograph or rendering as if it is the reality. There is no ebb and flow of depth and surface, nor invitation to see *into*; rather, we look *at* digital images. Seurat sought out the paper's surface texture as part of the work; his "points" are always large enough to be active components of the visual process. Digital technology has evolved in the opposite direction, aspiring

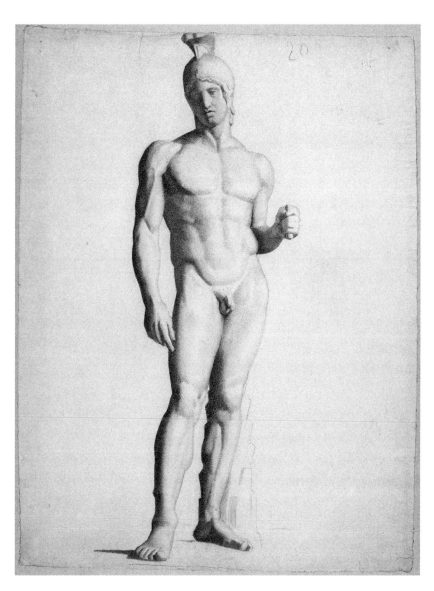

12.8

Warrior with Helmet, Copy after the Borghese "Ares," Georges Seurat, c. 1877–1878. Photograph © 2019 Museum of Fine Arts, Boston.

to make pixels disappear, regarding them as a hindrance to seeing. Hence my referring to digital images as hallucinatory.

Hallucination is a mental state that displaces the reality our senses give us for a reality dominated by a state of mind. As pixel counts reach into the millions, we effectively are convincing ourselves that the two-dimensional surface we are viewing is complete, leaving nothing to the imagination—or perhaps nothing *for* the imagination. Soon we might find ourselves believing the screen in front of us *is* reality.

SENSIBLE AND SUSTAINABLE

In the design process, materials influence the imagination and the designs that follow. The things and environments we design—clothing, furniture, buildings, cars, pavement, plantings, signage, all extensions and enactments of our self-world—influence how we interact with our surroundings. They are what we touch, experience, and form as our individual *Umwelt*, the sum of all that potentially integrates with and determines our sense of self. We've transformed the natural world and remade it by design and circumstance, yet we still apprehend our *Umwelt* as a normalized, seamless continuum, neither artificial nor natural. What has changed is our interaction with things.

Most modern city dwellers spend little or no time materially manipulating and transforming their physical surroundings, at least in comparison to, say, farmers, who have a far deeper and comprehensive understanding of the symbiosis of their actions and their environment. Farmers are attentive *because* they are touching and feeling and working with the material at hand, both living and inert, while having thoughts about and making decisions guided by material's reaction. Meanwhile, the linear methods of industrialization—now ensconced as the default approach to production—have blithely ignored consequences to a point where the utter depletion and destruction of life and resources has generated a resistance so immense it can no longer be ignored. This has happened in part because the linearity of production is missing a crucial feature: the *circularity* that drives creativity.

Reciprocity, attentiveness, action leading to thought, sensibility, and empathy all bend the arc of enactment back around to itself, whether at the scale of the individual or society. Sustainability, a now ubiquitous and overused term, is the attempt to impart *circularity* by other means, without the sensibility of the farmer or craftsman—whose work was sustainable long before the word was coined. In the contemporary version, sustainability works to recreate *sensibility* by technical and political means. Experience and sensibility are being replaced by data, the web is slowly turning into a planetary nervous system, and designers are turning their efforts toward interactive devices, mass customization, touch-responsive hardware, and personalized environments. It has become more and more common to see museumgoers spending time in digital interactions with interactive information kiosks rather than engaging with actual works on display, a fitting analogue to the way "sustainable" practices have displaced sensible practices.

We are not about to return to an agrarian society, and harnessing digital technology to build and advance a sustainable approach to resource use needs no defending. The point here is to recognize that the immersive nature of our digital reality profoundly affects our being-in-the-world. The feedback of our physical environment is integral to our mind, but paradoxically there is little awareness of it because of the persistent assumption that thought originates in the mind and is unaffected by its physical context. It is ironic that software strives to eliminate resistance while recreating it with haptic feedback mechanisms in order to harness our physical cognition. As these trends continue, we will need to find ways to integrate the mental leverage of the digital universe with the creative and imaginative empowerment of our physical and material environment.

DIGITIZATION'S IMPACT ON THE MIND

Does shifting from a physical experience to its digital equivalent affect the mind? Consider books and the transition from printed page to digital display. A reasonable assumption—namely, that reading from a screen should be the equivalent of reading the same text from a printed page—seems to be confirmed by the rise of electronic print and e-books. Nevertheless, physical printing is thriving and recent e-books sales are flat, even declining.[10] This isn't proof of anything, but it does suggest that material experience cannot be easily disassociated from our mind.

Another example compares hand drafting to computer-aided drafting (CAD). For a brief time at the start of this century, architecture students at the college where I teach—already thoroughly trained in hand drafting techniques—were taught the new CAD along with beginning students who had very little experience with hand drafting. CAD work is all keyboard and mouse, requiring no more dexterity than that of a typist. Although both sets of students learned the various commands at a similar pace, it became clear that those students with strong hand drafting abilities had a far deeper understanding of what they were doing. They were able to work tacitly, bringing the subtle material sensibilities of hand drafting (line weights, pencil hardness, implied spatial depth, etc.) to bear on their digital work. Moreover, although they were just typing and clicking, this set of students operated in an enactive fashion: as the drawing came into existence, they could feel and make judgments as if they were working with a hand drawing. In contrast, the beginning architecture students, although having as much ability in manipulating the various CAD commands, struggled to make a drawing amount to more than the dull accumulation of technical steps. They could not find the intelligence of their hand in the CAD process, an intelligence that could immediately be recognized in the CAD drawings of the students with hand drawing experience.

Does digitization affect our imagination? In addition to witnessing the transformation of student work from being hand drawn to being produced almost exclusively by digital means, I have reviewed the portfolios of hundreds of college applicants from around the world and have seen firsthand the profound effect embracing digital software and hardware has had on their work. Submitted portfolios are increasingly difficult to characterize or place, because digitally rendered projects are often more readily identified by the software employed than by the designer who used that software. Putting aside the influences of social media as the progenitor of an emerging like-mindedness that turns individual into typical work, what is consistent about digital work is how difficult it is as a distinguishing medium. Whereas a hand drawing is like a signature, a digital drawing or artifact tends toward anonymity. This is striking when one considers the enormous, almost infinite range and potential of digital processes. Equally striking is the restriction of activities that software, by its sequential logic, tends to instill. The inherently linear procedures amplify the narrowness of goal orientation: the question as one learns software is always "how do I do this?" As software programs are mastered, the more important question— "what is possible?"—is difficult to pull into the realm of chance or intuitive impulses, since breaking the chain of procedural commands often yields incoherence. I am reminded of the automated cooking equipment in chain restaurants: what is designed to take the "guesswork" out of the process at the same time cancels out the freedom afforded by engaging in the culinary arts. If exchanging material-based media for digital processes has consequences, a narrowing of the imagination would surely be a sign of trouble, particularly in design and the arts, which celebrate the human capacity for visual imagination.

Digitization's impact on the brain is the subject of innumerable articles and debates. "Digital dementia," "internet use disorder," "cognitive overload," and "nomophobia" (a name proposed to describe the fear of being out of cellular phone contact) are some of the new maladies said to result from the overuse of and overdependence on digital technologies. Framed in the terms of neurology, medicine, and psychology, these are characterized as addictions of the vulnerable neuroplastic brain requiring the cure of withdrawal. On the other side of the argument is the optimistic contention that humans are becoming more intelligent because the digital extension and networking of the brain has freed us from having to remember or know information as long as we are aware of where to get it. In both views, the brain is conceived as a functional organ, parallel to and thus vulnerable to or happily primed to join with digital technology, reinforcing Ray Kurzweil's idea of the brain as a computational machine or pattern recognition device.[11]

But a brain is not a mind, and the mind is not a brain. The mind, as the examples throughout this book show, is comprised of the interactions of the brain/sensorimotor system and body with its surrounding physical environment. In the context of the Clark and Chalmers extended mind hypothesis,[12] a pencil and notepad or a digital tablet are equivalent functional and cognitive extensions of mind, just as the 20,000-year-old Ishango bone was in its day and as a handheld calculator continues to be today. If so, then the argument can be made that digital technology is natural for our species and should be embraced as such.

However, there is a substantial difference between virtual and physical action. The mind's interactions, recursive, reiterative, reactive and enactive, flow deeply through the physical, material environment and the body and brain/sensorimotor system. Digitization extends the mind, but it also disengages the mind. Touching a device or clicking a mouse to manipulate software on a screen redirects the flow away from the very physicality that undergirds the cognitive and thinking process of our species.

Our evolutionary dependence on material and the physical environment to enact our minds was and remains a powerful lever that makes it possible for us to adapt to and fuse with almost every environment we have encountered. Are we aware, though, of how much we are *part* of that environment and how much of our mind is enacted with and through our physical interactions with the *Umwelt*? The very seamlessness of the enactive mind might prompt such a lack of awareness, making the effects of digitization difficult to perceive or judge—although there is ready evidence to suggest we ignore these effects at our peril.

Solitary confinement, one of the cruelest of punishments humans have devised, is based on removing any interaction or contact with one's social and physical environment beyond the stark cell, and has been shown to have severely negative effects. Acute reactive prison psychosis is a clinical term for the resulting breakdown of the mind, which includes symptoms such as delirium, loss of boundary (between self-sense and environment), hallucination, withdrawal, panic, and anxiety.[13] The 1950s extended the horror, giving us "brainwashing," in which prisoners are subjected to severe limitations in their sensory environment in order to manipulate or "condition" their minds. That solitary confinement continues as a practice is further evidence of a deep-rooted and misplaced assumption that the mind can be held apart from its physical environment.

If we look back at the psychological wellbeing of drone pilots in the Las Vegas desert (discussed in chapter 8), or the imagination of art students with their unimaginative apple drawings (discussed in chapter 4), or the unconsidered empathy of the theatergoer elicited by tearjerker productions Brecht so decried (discussed in chapter 7), we can recognize that each of these people lacked an element of *resistance* in their *Umwelt*, without which an enactive mind, like a malnourished body, is diminished. These examples help differentiate between resistance and stimulation. The latter term is generally used to describe things that activate the mind, but here I mean the overt, gratuitous kind that comes at you almost incessantly—for instance, as you listen to very loud music. Our drone pilot, art students, and early twentieth-century German theatergoers didn't lack for stimulation, they lacked for the very resistance that digital technology—though hugely stimulating—also lacks, and even seeks to do away with.

Stimulation is a prodding, a goading, a zapping that flows in one direction, placing the body and sensorimotor system into a more passive mode, receiving more and doing less. It also removes authorship: being stimulated is like following an instruction, which is not the same as acting on one's own. Resistance, on the contrary, is a latent potential that quietly awaits. If our physical *Umwelt* can be understood this way, as a potential that also includes affordances, waiting to participate, it leads naturally to physical action initiated and authored by the self, which in turn generates a resistance specific to the action, and a new action that responds to resistance, and so on, action and resistance flowing back and forth and thus catalyzing the mind's enaction; situated thinking, empathy, and imagination are but three of the essential qualities of mind emerging from and contingent on such a reciprocity.

This book began with a proposition: that the hunches, intuitions, realizations, insights, and understanding experienced by those working with material should be considered a form of thought, even if such "soft thought" is more tied to feelings and empathic sensibility. The very reality of soft thinking challenges the premise that thinking is detached from emotion. Why should feeling be opposed to thinking, instead of part of a continuum that includes both? Thinking uses language abstractly and self-reflectively and, in doing so, gains autonomy from the senses, or so it seems. But as the examples in this book reveal, representations, experience, and the influence of physical materials are never at a remove from thought, and, when actively engaging the *Umwelt*, we are in fact deeply thinking from, with, and through it. Tacit thinking, visual thinking, situated thinking, spatial thinking, imagination, and thinking with the hands—each and every one a soft thought process—all result from the body and its senses immersed in physical reality. They involve feeling as part of thinking, contradicting the traditional view of thought and feeling as a duality.

If all thought, including soft thought, can be understood as not only influenced by physical reality but also structured by it, the location of thought begins to align with that of the enactive mind, between brain matter and the *Umwelt*. Thinking outside the box, it turns out, is thinking outside the brain.

The action/resistance-based enactive model puts our hands in the middle of thinking, between the brain and physical reality. Hands are brainlike, extensions of the brain tied to it neurologically. Hands are also material-like, the extension of material properties entering the mind. We need to remind ourselves that all making—which is, after all, a great and complex accumulation of thinking processes—arises from the dexterity of our hands. Even the most automated silicon chip factory has its roots in hands working with material, hands that make the parts that assemble into greater and greater complexities and machines, slowly drawing on all the dimensions of thinking. How we use our hands is not simply about tendons and muscles, but is a question of the mind.

This raises issues of education. Can we learn without the involvement of our hands and their participation in our physical material and *Umwelt*? If we shackle the hand—our zoologically determined apparatus—to the keyboard or touch screen, are we forgetting our hand's need for a coupled relation with its environment? What will happen to soft thinking in the face of disembodied online learning and online interaction that, despite efforts at feedback, has students follow a linear and prescribed set of instructions and learning modules?

Can our brain adjust to thrive in a digital world? Is it essential for the mind to enact itself in the resistance of the physical world? These are key questions of our time, and return us to where we began as a species. The completeness of early humankind's being-in-the-world was slowly exchanged for consciousness, dexterity, and sociability. Even as we were empowered by our new abilities, we continued to search for the lost immersion in the cosmic order, first with religion and then by reshaping the world into one of our making. What has remained unchanged is our physiological equipment, including the brain, which by its nature must engage with the resistance of material and flux of the *Umwelt* to enact the mind and develop empathy, imagination, and soft thinking. The greater the complexity and richness of engagement, the deeper the enactment, and the greater the connection to the world.

Jerome Bruner's definition of meaningful as a sense of connection and relevance that is greater than self perfectly describes being-in-the-world.[14] As we continue to immerse ourselves in a digital universe, we must ask ourselves: Where will we find meaning in our lives?

Notes

INTRODUCTION

1. Proust, *In Search of Lost Time*, 916.

2. Clark and Chalmers, "The Extended Mind."

3. Brown, *Arthur Mervyn*, 98.

CHAPTER 1: THINK OR DO

1. Wilson, *The Hand*.

2. Taylor, "Materiality," 302.

3. Hugo, *Notre-Dame of Paris*, 187.

4. Wayland, "Clay Modelers Shape the Future of Auto Design."

5. Ford Europe, "The Tradition of Clay Modelling in Car Design."

6. Bayley, *Harley Earl and the Dream Machine*, 38.

7. "Harley Earl, Stylist Who Made Autos Long, and Low, Dies at 75," *New York Times,* April 11, 1969.

8. Bergson, *Matter and Memory*.

9. Merleau-Ponty, *Phenomenology of Perception*, 370.

10. Fodor, *Representations*.

11. Clark and Chalmers, "The Extended Mind."

12. Ibid.

13. Varela, Thompson, and Rosch, *The Embodied Mind*.

14. Leroi-Gourhan, *Gesture and Speech*.

15. Malafouris, *How Things Shape the Mind*, 117.

16. Fadiga and Craighero, "Hand Actions and Speech Representation in Broca's Area."

17. Wegner, "The Mind's Best Trick."

18. Harris, *Free Will*.

19. Dewey, *The Later Works*, 124.

20. Thomas, *Oeuvres*, 31.

21. Keohane, "How Facts Backfire."

22. Virgil, *Eclogues* X.

23. Kant, *Critique of Judgment*.

CHAPTER 2: IN THE BEGINNING

1. *Merriam-Webster Dictionary*, 356.

2. Aristotle, "De anima," in *Basic Works*.

3. Bergson, *Matter and Memory*, 106.

4. Loftus, "Our Changeable Memories."

5. Bruner, *Acts of Meaning*, 33.

6. Aubert et al., "Pleistocene Cave Art from Sulawesi, Indonesia."

7. Clottes and Lewis-Williams, *The Shamans of Prehistory*, 85.

8. Halverson, "The First Pictures," 396.

9. Ibid.

10. Ibid.

11. Merleau-Ponty, "Eye and Mind," 78–79.

12. Malafouris, *How Things Shape the Mind*, 204.

13. William Padula, neuro-optometrist, multiple conversations with the author, 2013–2016.

14. Cohen and Bennett, "Why Can't Most People Draw What They See?"

15. Heinrich, *Racing the Antelope*, 174–179.

16. Gould, "The Tallest Tale," my emphases.

17. Stout et al., "Neural Correlates of Early Stone Age Toolmaking."

18. Studdert-Kennedy and Louis Goldstein, "Launching Language."

19. Hodgson, "Evolution of the Visual Cortex."

20. Donald, *Origins of the Modern Mind*, 151.

21. Ibid., 152.

22. Gould, *The Structure of Evolutionary Theory*, 745.

23. Harari, *Sapiens*, 28.

24. Jung, *The Archetypes and the Collective Unconscious*, 42.

25. Shelley, "A Defence of Poetry," 4.

CHAPTER 3: WORDS MATTER

1. Leroi-Gourhan, *Gesture and Speech*.

2. Ramachandran and Hubbard, "Neurocognitive Mechanisms of Synesthesia."

3. Ramachandran, *The Tell-Tale Brain*, 182.

4. Ibid., 108.

5. Robson, "Kiki or Bouba?"

6. Gould, *Ontogeny and Phylogeny*, 1.

7. Kazakov and Tsoulas, "Applying Recapitulation Theory to Language."

8. Stern and Stern, *Psychology of Early Childhood*," 180.

9. Vygotsky, *Thought and Language*, 65.

10. Ibid.

11. Kohlberg, Yaeger, and Hjertholm, "Private Speech."

12. Gibson, *The Senses Considered as Perceptual Systems*.

13. Lawler, "Writing Gets a Rewrite."

14. Mankiewicz, *The Story of Mathematics*, 10.

15. Schmandt-Besserat, *Before Writing*.

CHAPTER 4: REPRESENTING

1. Pound, *In a Station of the Metro*, 12.

2. Pound, "Vorticism."

3. Schjeldahl, "Improvising in Art and Life."

4. Kieras, "Beyond Pictures and Words."

5. Lakoff and Johnson, *Philosophy in the Flesh*, 26.

6. Husserl, "The Thesis of the Natural Standpoint and Its Suspension," 114.

7. Van Gogh, "Letter . . . to Theo c. 12–18 December 1882."

8. Geary, *I Is an Other*, 142.

9. Smith, *The Century Dictionary*, 5428.

10. Sayce, *Introduction to the Science of Language*, 180–181.

11. Emerson, "The Poet."

12. Gass, "Representation and the War for Reality."

13. Discussion with the author, Harvard University Graduate School of Design seminar, 1986. See also Evans, *Translations from Drawing to Building and Other Essays*, 165–166.

14. Graziano, "Are We Really Conscious?"

15. Dickens, *Hard Times*.

16. Kurzweil, *How to Create a Mind*, 6.

17. Buckner and Krienen, "The Evolution of Distributed Association Networks in the Human Brain."

18. Ibid., 658.

19. Ramachandran, *The Tell-Tale Brain*.

20. Bachelard, *On Poetic Imagination and Reverie*, 14.

21. Ibid.

22. Brodsky, *Watermark*, 129.

23. Bachelard, *On Poetic Imagination and Reverie*.

24. Ericsson, "The Acquisition of Expert Performance," 22.

25. Stevens, *Opus Posthumous*, 195.

26. Pound, "A Retrospect," 6.

27. Poe, *The Narrative of Arthur Gordon Pym of Nantucket*, 155.

28. Kiš, *Garden, Ashes*, 24.

29. Shelley, "A Defence of Poetry."

30. Wilde, *The Picture of Dorian Gray*, 22.

31. Galileo, *Discoveries and Opinions of Galileo*, 92.

32. Gemünden, *Framed Visions*, 1.

33. Shelley, "A Defence of Poetry."

34. Bachelard, *On Poetic Imagination and Reverie*.

CHAPTER 5: THE IDEA AND THE ACTUAL

1. Plato, *The Collected Dialogues of Plato*.

2. Aristotle, *The Basic Works of Aristotle*.

3. Descartes, *Meditations on First Philosophy*, 21.

4. Copernicus, *De revolutionibus orbium coelestium*, 405.

5. Kepler, *Mysterium cosmographicum*.

6. Sagan, *Cosmos*, 52.

7. Bacon, *Novum organum*, 20.

8. Lévi-Strauss, "The Science of the Concrete."

9. Gilges, "Some African Poison Plants and Medicines of Northern Rhodesia"; Conklin, "The Relation of Hanunóo Culture to the Plant World"; Speck, "Reptile Lore of the Northern Indians."

10. Lévi-Strauss, "The Science of the Concrete," 10.

CHAPTER 6: PROJECTION

1. Nietzsche, *The Birth of Tragedy*.

2. Damen, "Classical Drama and Society."

3. Pérez-Gómez and Pelletier, *Architectural Representation and the Perspective Hinge*, 10.

4. Whitman, "A Song of Occupations."

5. Pliny, *Natural History*, 373.

6. Mallarmé, "Hérodiade," 25.

7. Bachelard, *On Poetic Imagination and Reverie*.

8. Piaget and Inhelder, *The Child's Conception of Space*, 203.

9. Duckworth, *Science Education*, 143.

10. Kant, *Critique of Pure Reason*.

11. Manetti, *The Life of Brunelleschi*.

12. Alberti, *On Painting*, 17.

13. Piero della Francesca, *De prospectiva pingendi*, 17.

14. Białostocki, *The Message of Images*, 86.

15. Ibid.

16. Jenison, *Tim's Vermeer*.

17. Mach, *Popular Scientific Lectures*, 76.

18. Florensky, *Beyond Vision*, 262.

19. Ibid.

20. Mach, *The Analysis of Sensations and the Relation of the Physical and the Psychical*, 10.

21. Aristotle, *Poetics*, 19, 1449a.

CHAPTER 7: THE TRANSMITTING HAND

1. Rich, *The Material and Methods of Sculpture*, 262.

2. Ibid., 265.

3. Ibid.

4. Payne, "The Sculptor-Architect's Drawing and Exchanges between the Arts."

5. Foote, "Extracting Desire," 35.

6. Ibid., 31.

7. Gans, *The Le Corbusier Guide*, 176.

8. Pauly, "The Chapel of Ronchamp as an Example of Le Corbusier's Creative Process," 128.

9. Evans, *The Projective Cast*, 295.

10. Klee, *Pedagogical Sketchbook*.

11. Soltan, "Working with Le Corbusier," 6.

12. Le Corbusier, *Le poème de l'angle droit*.

13. Williams and Bargh, "Experiencing Physical Warmth Promotes Interpersonal Warmth."

14. Blake, *Complete Writings*, 818.

15. Blake, letter to Thomas Butt, November 22, 1802, in *The Letters of William Blake*, 000.

16. Ford Europe, "The Tradition of Clay Modelling in Car Design."

17. Proust, *In Search of Lost Time*.

18. Heidegger, "Conversation on a Country Path about Thinking."

19. Stirling, "Ronchamp," 161.

20. LaCapra, *History in Transit*, 135.

21. Ibid.

CHAPTER 8: WORKING WITH MATERIAL

1. Examples can be seen at George Nakashima, Woodworker: http://www.nakashimawoodworker.com.

2. Examples can be seen at Artsy, "Giuseppe Penone," https://www.artsy.net/artist/giuseppe-penone; further examples can be seen at Art Gallery of Toronto, "Giuseppe Penone: The Hidden Life Within," http://www.ago.net/giuseppe-penone-the-hidden-life-within.

3. Lesser, *You Say to Brick*, 5.

4. Gallwey, *The Inner Game of Tennis*, 11.

5. Bachelard, *On Poetic Imagination and Reverie*, 68.

6. Rowe, *As I Was Saying*, 356.

7.   Retica, "Drone-Pilot Burnout."

8.   Coriolis, "Sur les équations du mouvement relatif des systèmes de corps."

9.   Nietzsche, *The Birth of Tragedy*, 22.

10. Murphy and Roberts, *Dialectic of Romanticism*, 4.

11. Evans, *Translations from Drawing to Building and Other Essays*, 270 (my emphasis).

12. Bergson, *An Introduction to Metaphysics*, 9.

13. Hammer-Tugendhat, Hammer, and Tegethoff, *Tugendhat House*, 20.

14. Klee, *The Diaries of Paul Klee*, 8.

15. Hammer-Tugendhat, Hammer, and Tegethoff, *Tugendhat House*.

16. Tugendhat, "Letter in Response to Walter Riezler's Article."

17. Ibid.

18. Clottes and Lewis-Williams, *The Shamans of Prehistory*.

CHAPTER 9: MATERIAL AND UMWELT

1.   Benjamin, *Berliner Kindheit um neunzehnhundert*, translation by the author.

2.   Kalevi, "Umwelt."

3.   Uexküll, *Theoretical Biology*.

4.   Heidegger, *Being and Time*.

5.   Clark and Chalmers, "The Extended Mind"; Varela, Thompson, and Rosch, *The Embodied Mind*.

6.   Gadamer, *Truth and Method*.

7.   Bergson, *Matter and Memory*, 106.

8.   Rossi, *A Scientific Autobiography*, 19.

9.   Gadamer, *Truth and Method*.

10. Hutchins, "How a Cockpit Remembers Its Speeds."

11. Ibid., 284.

12. Olaf Herrmann, conversation with the author, July 3, 2018.

13. Heidegger, "The Origin of the Work of Art."

14. Ibid.

15. Harari, *Sapiens*.

16. Dean, *A Culture of Stone*, 67.

17. Bennett, *Vibrant Matter*, viii (emphasis in the original).

CHAPTER 10: INSIDE THE DESIGN PROCESS

1. Lévi-Strauss, "The Science of the Concrete," 16.

2. Ibid.

3. Le Corbusier, *Towards a New Architecture*, 11.

4. Lévi-Strauss, "The Science of the Concrete," 18.

5. Duncker, "On Problem-Solving."

6. Ibid.

7. Brown, *Penicillin Man*, 2.

8. Kant, *Critique of Pure Reason*, 193.

9. Hughes, *The Architects*.

10. Kaidanovsky, Solonitsyn, and Freyndlikh, *Stalker*.

11. Bachelard, *On Poetic Imagination and Reverie*, 134.

12. Gass, "Representation and the War for Reality," 96.

13. Lévi-Strauss, "The Science of the Concrete," 14.

14. Kant, *Critique of Pure Reason*, 45.

15. Kahn, *Essential Texts*, 69.

16. Pallasmaa, *The Eyes of the Skin*, 30–31.

CHAPTER 11: DESIGN TOOLS AND THEIR ROLES

1. Valéry, "The Idea of Art," 25 (emphasis in the original).

2. Tulving, "Concepts of Human Memory," 12.

3. Serra, "Verb List."

4. Gould, "One Man's Approach to a Basic Course in Geological Sciences," 121.

5. Merleau-Ponty, "Eye and Mind," 55.

6. Gould, "One Man's Approach to a Basic Course in Geological Sciences," 121.

7. Flusser, *Towards a Philosophy of Photography*, 15.

8. Sacks, "Neurology and the Soul."

9. Gabler, "The Elusive Big Idea."

10. Gass, "Representation and the War for Reality," 77.

11. Andreas-Salomé, *Looking Back: Memoirs*, 20.

12. Klee, "Creative Credo."

13. Schelling, *The Philosophy of Art*, 32 (emphasis in the original).

CHAPTER 12: SMOOTHNESS

1. Aristotle, *Physics*, in *Basic Works*.

2. Aristotle, *Metaphysics*, in *Basic Works*.

3. Heidegger, "The Question Concerning Technology," 313.

4. Benjamin, "The Work of Art in the Age of Mechanical Reproduction."

5. Hui, "Form and Relation."

6. Ibid.

7. Gilbert, *Du mode d'existence des objets techniques*.

8. Conrad, *Die beginnende Schizophrenie*; this translation from Hubscher, "Apophenia."

9. Maslow, *The Psychology of Science*. The original version of this quote, presented above in its vernacular form, is: "I suppose it is tempting, if the only tool you have is a hammer, to treat everything as if it were a nail."

10. Ivana Kottasová, "Real Books Are Back. E-book Sales Plunge Nearly 20%," CNN, April 27, 2017, http://money.cnn.com/2017/04/27/media/ebooks-sales-real-books/.

11. Kurzweil, *How to Create a Mind*.

12. Clark and Chalmers, "The Extended Mind."

13. Grassian, "Psychiatric Effects of Solitary Confinement."

14. Bruner, *Acts of Meaning*.

Bibliography

Alberti, Leon Battista. *On Painting*. Translated by Rocco Sinisgalli. Cambridge, UK: Cambridge University Press, 2011. First published in Italian in 1435–1436 as *Della pittura*, and then in a more technical version in Latin in 1439–1441 as *De pictura*.

Andreas-Salomé, Lou. *Looking Back: Memoirs*. Translated by Breon Mitchell. New York: Paragon House, 1990. First published posthumously in 1951 as *Lebensrückblick: Grundriss einiger Lebenserinnerungen*.

Aristotle. *De anima*. Translated by C. D. C. Reeve. Cambridge, MA: Hackett, 2017).

Aristotle. *The Basic Works of Aristotle*. Edited by Richard McKeon. Various translators. New York: Random House, 1941.

Aristotle. *Poetics of Aristotle*. Edited and translated by S. H. Butcher. New York: Macmillan, 1902.

Aubert, Maxime, Adam Brumm, Muhammad Ramli, Thomas Sutikna, E. Wahyu Saptomo, Budianto Hakim, MJ vd Morwood, Gerrit D. van den Bergh, Leslie Kinsley, and Anthony Dosseto. "Pleistocene Cave Art from Sulawesi, Indonesia." *Nature* 514, no. 7521 (2014): 223–227.

Bachelard, Gaston. *On Poetic Imagination and Reverie*. Translated by Colette Gaudin. Thompson, CT: Spring Publications, 2014.

Bacon, Francis. *Novum organum*. Edited by Joseph Devey. Translated by William Wood. New York: P. F. Collier & Son, 1902. First published in 1620.

Bambach, Carmen C. *Michelangelo: Divine Draftsman and Designer*. New York: Metropolitan Museum of Art, 2017.

Bayley, Stephen. *Harley Earl and the Dream Machine*. New York: Alfred A. Knopf, 1983.

Benjamin, Walter. *Berliner Kindheit um neunzehnhundert*. Frankfurt: Suhrkamp Verlag, 1950.

Benjamin, Walter. "The Work of Art in the Age of Mechanical Reproduction." In *Walter Benjamin and Art*. Edited by Andrew Benjamin. London: Bloomsbury Academic, 2005. First published in 1935 as "Das Kunstwerk im Zeitalter seiner technischen Reproduzierbarkeit."

Bennett, Jane. *Vibrant Matter: A Political Ecology of Things*. Durham, NC: Duke University Press, 2009.

Bergson, Henri. *An Introduction to Metaphysics*. Translated by T. E. Hulme. New York: G. P. Putnam's Sons, 1912.

Bergson, Henri. *Matter and Memory*. Translated by Nancy Margaret Paul and W. Scott Palmer. New York: Zone Books, 1990. First published as *Matière et mémoire: Essai sur la relation du corps à l'esprit*, 1896.

Białostocki, Jan. *The Message of Images: Studies in the History of Art*. Vienna: Irsa Verlag, 1988.

Blake, William. *Complete Writings*. Edited by Geoffrey Keynes. Oxford, UK: Oxford University Press, 1972.

Blake, William. *The Letters of William Blake*. Edited by Geoffrey Keynes. London: Hart-Davis, 1956.

Brodsky, Joseph. *Watermark*. New York: Farrar, Straus & Giroux, 1992.

Brown, Charles Brockden. *Arthur Mervyn; or Memoirs of the Year 1793*. Philadelphia: David McKay, 1887.

Brown, Kevin. *Penicillin Man: Alexander Fleming and the Antibiotic Revolution*. Gloucestershire, UK: History Press, 2013.

Bruner, Jerome. *Acts of Meaning*. Cambridge, MA: Harvard University Press, 1990.

Buckner, Randy L., and Fenna M. Krienen, "The Evolution of Distributed Association Networks in the Human Brain." *Trends in Cognitive Sciences* 17, no. 12 (2013): 648–665.

Clark, Andy, and David J. Chalmers. "The Extended Mind." *Analysis* 58, no. 1 (1998): 7–19.

Clottes, Jean, and David Lewis-Williams. *The Shamans of Prehistory*. Translated by Sophie Hawkes. New York: Harry N. Abrams, 1996.

Cohen, Dale J., and Susan Bennett. "Why Can't Most People Draw What They See?" *Journal of Experimental Psychology: Human Perception and Performance* 23, no. 3 (1997): 609–621.

Conklin, Harold C. "The Relation of Hanunóo Culture to the Plant World." PhD dissertation, Yale University, 1954.

Conrad, Karl. *Die beginnende Schizophrenie. Versuch einer Gestaltanalyse des Wahns*. Stuttgart: Thieme, 1958.

Copernicus, Nicolaus. *De revolutionibus orbium coelestium*. Nuremberg: Johannes Petreius, 1543.

Coriolis, Gaspard-Gustave de. "Sur les équations du mouvement relatif des systèmes de corps." *Journal de l'Ecole royale polytechnique* 15 (1835): 144–154.

Damen, Mark. "Classical Drama and Society." Course materials, Utah State University, Logan, Utah, 2012. https://www.usu.edu/markdamen/ClasDram/chapters/041gkorig.htm.

Dean, Carolyn J. *A Culture of Stone: Inka Perspectives on Rock*. Durham, NC: Duke University Press, 2010.

Descartes, René. *Meditations on First Philosophy*. Translated and edited by John Cottingham. Cambridge, UK: Cambridge University Press, 1996. First published in 1641 as *Meditationes de prima philosophia, in qua Dei existentia et animæ immortalitas demonstratur.*

Dewey, John. *The Later Works, 1925–1953*. Volume 8: 1933. Carbondale: Southern Illinois University Press, 1986.

Dickens, Charles. *Hard Times*. In *The Shorter Novels of Charles Dickens*. London: Wordsworth Editions, 2005.

Doidge, Norman. *The Brain That Changes Itself: Stories of Personal Triumph from the Frontiers of Brain Science*. New York: Penguin, 2007.

Donald, Merlin. *Origins of the Modern Mind: Three Stages in the Evolution of Culture and Cognition*. Cambridge, MA: Harvard University Press, 1991.

Duckworth, Eleanor. *Science Education: A Minds-On Approach for the Elementary Years*. Hillsdale, NJ: Erlbaum Associates, 1990.

Duncker, Karl. "On Problem-Solving." *Psychological Monographs* 58, no. 5 (1945).

Emerson, Ralph Waldo. "The Poet." In *The Works of Ralph Waldo Emerson*. Vol. 3, *Essays*. Boston and New York, 1909.

Ericsson, K. Anders. "The Acquisition of Expert Performance: An Introduction to Some of the Issues." In *The Road to Excellence: The Acquisition of Expert Performance in the Arts and Sciences, Sports, and Games*. Edited by K. Anders Ericsson. New York: Psychology Press, 2014.

Evans, Robin. *The Projective Cast: Architecture and Its Three Geometries*. Cambridge, MA: MIT Press, 1995.

Evans, Robin. *Translations from Drawing to Building and Other Essays*. London: Architectural Association Publications, 2011.

Fadiga, Luciano, and Laila Craighero. "Hand Actions and Speech Representation in Broca's Area." *Cortex* 42, no. 4 (June 2006): 486–490.

Florensky, Pavel. *Beyond Vision: Essays on the Perception of Art*. Translated by Wendy Salmond. London: Reaktion Books, 2002.

Flusser, Vilém. *Towards a Philosophy of Photography*. London: Reaktion Books, 2000.

Fodor, Jerry A. *Representations: Philosophical Essays on the Foundations of Cognitive Science*. Cambridge, MA: MIT Press, 1981.

Foote, Jonathan. "Extracting Desire: Michelangelo and the 'forza di levare' as an Architectural Premise." In *The Material Imagination: Reveries on Architecture and Matter*. Edited by Matthew Mindrup. New York: Routledge, 2016.

Ford Europe. "The Tradition of Clay Modelling in Car Design." YouTube video, 02:18. Published April 8, 2014. https://www.youtube.com/watch?v=uAjFedlj2rE.

Gabler, Neal. "The Elusive Big Idea." *New York Times*, August 13, 2011.

Gadamer, Hans-Georg. *Truth and Method.* Translated by J. Weinsheimer and D. G. Marshall. New York: Continuum, 1989. Originally published in 1960 as *Wahrheit und Methode.*

Galileo Galilei. *Discoveries and Opinions of Galileo.* Translated by Stillman Drake. New York: Anchor Books, 1957.

Gallwey, W. Timothy. *The Inner Game of Tennis.* New York: Random House, 1974.

Gans, Deborah. *The Le Corbusier Guide.* New York: Princeton Architectural Press, 1987.

Gass, William H. "Representation and the War for Reality." *Salmagundi* 55 (1982): 72.

Geary, James. *I Is an Other: The Secret Life of Metaphor and How It Shapes the Way We See the World.* New York: Harper Perennial, 2011.

Gemünden, Gerd. *Framed Visions: Popular Culture, Americanization, and the Contemporary German and Austrian Imagination.* Ann Arbor, MI: University of Michigan Press, 1998.

Gibson, James J. *The Senses Considered as Perceptual Systems.* Boston: Houghton Mifflin, 1966.

Gilges, Wilhelm. "Some African Poison Plants and Medicines of Northern Rhodesia." Occasional Papers, Rhodes-Livingstone Museum, no. 11, 1955.

Gould, Stephen Jay. "One Man's Approach to a Basic Course in Geological Sciences." *Journal of Geological Education* 32, no. 2 (1984): 120–122.

Gould, Stephen Jay. *Ontogeny and Phylogeny.* Cambridge, MA: Belknap Press of Harvard University Press, 1977.

Gould, Stephen Jay. *The Structure of Evolutionary Theory.* Cambridge, MA: Harvard University Press, 2002.

Gould, Stephen Jay. "The Tallest Tale." *Natural History* 105, no. 5 (1996): 18–24.

Grassian, Stuart. "Psychiatric Effects of Solitary Confinement." *Washington University Journal of Law and Policy* 22 (2006): 325.

Graziano, Michael S. A. "Are We Really Conscious?" *New York Times*, October 12, 2014.

Halverson, John. "The First Pictures: Perceptual Foundations of Paleolithic Art." *Perception* 21, no. 3 (1992): 396.

Hammer-Tugendhat, Daniela, Ivo Hammer, and Wolf Tegethoff. *Tugendhat House: Ludwig Mies van der Rohe.* Basel: Birkhäuser Verlag, 2015.

Harari, Yuval Noah. *Sapiens.* London: Vintage, 2014.

Harris, Sam. *Free Will.* New York: Free Press, 2012.

Heidegger, Martin. *Being and Time.* Translated by John Macquarrie and Edward Robinson. New York: Harper, 1962. First published in 1927 as *Sein und Zeit.*

Heidegger, Martin. "Conversation on a Country Path about Thinking." In *Discourse on Thinking: A Translation of Gelassenheit.* Translated by John M. Anderson and E. Hans Freund. New York: Harper & Row, 1966.

Heidegger, Martin. "The Origin of the Work of Art." In *Basic Writings*. Edited by David Farrell Krell, revised and expanded edition. San Francisco: HarperSanFrancisco, 1993.

Heidegger, Martin. "The Question Concerning Technology." In *Basic Writings*. Edited by David Farrell Krell, revised and expanded edition. San Francisco: HarperSanFrancisco, 1993.

Heinrich, Bernd. *Racing the Antelope: What Animals Can Teach Us about Running and Life*. New York: HarperCollins, 2001.

Hendrix, Lee, editor. *Noir: The Romance of Black in 19th-Century French Drawings and Prints*. Los Angeles: Getty Publications, 2016.

Hodgson, Derek. "Evolution of the Visual Cortex and the Emergence of Symmetry in the Acheulean Techno-complex." *Comptes Rendus Palevol* 8, no. 1 (2009): 93–97.

Hubscher, Sandra L. "Apophenia: Definition and Analysis." Digital Bits Skeptic, blog, November 4, 2007, https://archive.is/20130121151738/http://www.dbskeptic.com/2007/11/04/apophenia -definition-and-analysis/#selection-537.0-541.20

Hughes, Robert. *The Architects: Louis Kahn*. New York: New Word City, 2015.

Hugo, Victor. *Notre-Dame of Paris*. Translated by John Sturrock. Harmondsworth, UK: Penguin, 1978. First published in 1831 as *Notre-Dame de Paris*.

Hui, Yuk. "Form and Relation: Materialism on an Uncanny Stage." *Intellectica* 61 (2014): 105–121.

Husserl, Edmund. "The Thesis of the Natural Standpoint and Its Suspension." In *Phenomenology and Existentialism*, edited by Robert Solomon. Lanham, UK: Rowman & Littlefield, 1972.

Hutchins, Edwin. "How a Cockpit Remembers Its Speeds." *Cognitive Science* 19, no. 3 (1995): 265–288.

Jenison, Tim. *Tim's Vermeer*. Film. Directed by Teller. Sony Picture Classics, 2013.

Jung, Carl. *The Archetypes and the Collective Unconscious*. Translated by R. F. C. Hull. Princeton, NJ: Princeton University Press, 1980.

Kahn, Louis. *Essential Texts*. Edited by Robert Twombly. New York: W. W. Norton, 2003.

Kaidanovsky, Alexander, Anatoli Solonitsyn, and Alisa Freyndlikh. *Stalker*. Directed by Andrei Tarkovsky. Russia: Mosfilm, 1979.

Kalevi, Kull. "Umwelt." In *The Routledge Companion to Semiotics*. Edited by Paul Cobley. London: Routledge, 2010.

Kant, Immanuel. *Critique of Judgment*. Translated by James Creed Meredith. Oxford, UK: Oxford University Press, 2007. First published in 1790 as *Kritik der Urteilskraft*.

Kant, Immanuel. *Critique of Pure Reason*. Translated by Paul Guyer and Allan Wood. Cambridge, UK: Cambridge University Press, 1998.

Kazakov, Dimitar, and George Tsoulas. "Applying Recapitulation Theory to Language." Conference on Ways of Protolanguage, Torun, Poland, 2009: Book of Abstracts, 23.

Keohane, Joe. "How Facts Backfire: Researchers Discover a Surprising Threat to Democracy: Our Brains," *Boston Globe,* July 11, 2010.

Kepler, Johannes. *Mysterium cosmographicum.* Tübingen, 1596.

Kieras, David. "Beyond Pictures and Words: Alternative Information-Processing Models for Imagery Effect in Verbal Memory." *Psychological Bulletin* 85, no. 3 (1978): 532–554.

Kiš, Danilo. *Garden, Ashes.* Translated by William Hannaher. New York: Harcourt Brace Jovanovich, 1975. First published in 1965 as *Bašta, pepeo.*

Klee, Paul. "Creative Credo." In *Theories of Modern Art: A Source Book by Artists and Critics.* Edited by Herschel Browning Chipp, with contributions by Peter Selz and Joshua C. Taylor. Berkeley: University of California Press, 1968.

Klee, Paul. *The Diaries of Paul Klee: 1898–1918.* Edited by Felix Klee. Berkeley: University of California Press, 1968.

Klee, Paul. *Pedagogical Sketchbook.* New York: Praeger, 1953.

Kohlberg, Lawrence, Judy Yaeger, and Else Hjertholm. "Private Speech: Four Studies and a Review of Theories." *Child Development* 39, no. 3 (1968): 691–736.

Kurzweil, Ray. *How to Create a Mind: The Secret of Human Thought Revealed.* New York: Penguin Group, 2013.

LaCapra, Dominick. *History in Transit: Experience, Identity, Critical Theory.* Ithaca, NY: Cornell University Press, 2004.

Lakoff, George, and Mark Johnson. *Philosophy in the Flesh: The Embodied Mind and Its Challenges to Western Thought.* New York: Basic Books, 1999.

Lawler, Andrew. "Writing Gets a Rewrite." *Science* 292, no. 5526 (2001): 2418–2420.

Le Corbusier. *Le poème de l'angle droit.* Translated by Kenneth Hylton. Paris: Fondation Le Corbusier, 1989.

Le Corbusier. *Towards a New Architecture.* Translated by Frederick Etchells. London: J. Rodker, 1931.

Leroi-Gourhan, André. *Gesture and Speech.* Translated by Anna Bostock Berger. Cambridge, MA: MIT Press, 1993. First published in 1964 as *Le geste et la parole.*

Leski, Kyna. *The Storm of Creativity.* Cambridge, MA: MIT Press, 2015.

Lesser, Wendy. *You Say to Brick: The Life of Louis Kahn.* New York: Farrar, Straus & Giroux, 2017.

Lévi-Strauss, Claude. "The Science of the Concrete." In *The Savage Mind.* London: Weidenfeld and Nicolson, 1966. First published in 1962 as *La pensée sauvage.*

Loftus, Elizabeth. "Our Changeable Memories: Legal and Practical Implications." *Nature Reviews Neuroscience* 4, no. 3 (2003): 231.

Mach, Ernst. *The Analysis of Sensations and the Relation of the Physical and the Psychical.* Translated by C. M. Williams. Chicago: Open Court, 1914.

Mach, Ernst. *Popular Scientific Lectures*. Translated by Thomas J. McCormack. Chicago: Open Court, 1895.

Malafouris, Lambros. *How Things Shape the Mind*. Cambridge, MA: MIT Press, 2013.

Mallarmé, Stéphane. "Hérodiade." In *Collected Poems*. Translated by Henry Weinfield. Berkeley: University of California Press, 1994.

Manetti, Antonio. *The Life of Brunelleschi*. University Park: Pennsylvania State University Press, 1970. First published c. 1480.

Mankiewicz, Richard. *The Story of Mathematics*. Princeton, NJ: Princeton University Press, 2001.

Maslow, Abraham M. *The Psychology of Science*. New York: Harper and Row, 1966.

Merleau-Ponty, Maurice. "Eye and Mind." In *Aesthetics*. Edited by Harold Osborne. Oxford, UK: Oxford University Press, 1972. First published in 1961 as *L'Œil et l'esprit*.

Merleau-Ponty, Maurice. *Phenomenology of Perception*. Translated by Colin Smith. London: Routledge, 1962. First published in 1945 as *Phénoménologie de la perception*.

Merriam-Webster Dictionary. Martinsburg, WV: Merriam-Webster, 2016.

Murphy, Peter, and David Roberts. *Dialectic of Romanticism*. London: Continuum International, 2004.

Nietzsche, Friedrich. *The Birth of Tragedy and The Genealogy of Morals*. Translated by Francis Golffing. New York: Anchor Books, 1956. First published in 1872 as *Die Geburt der Tragödie aus dem Geiste der Musik*.

Pallasmaa, Juhani. *The Eyes of the Skin*. West Sussex, UK: John Wiley & Sons, 2005.

Pauly, Danièle. "The Chapel of Ronchamp as an Example of Le Corbusier's Creative Process." In *Le Corbusier*. Edited by H. Allen Brooks. Princeton, NJ: Princeton University Press, 1987.

Payne, Alina. "The Sculptor-Architect's Drawing and Exchanges between the Arts." In *Donatello, Michelangelo, Cellini: Sculptors' Drawings from Renaissance Italy*. Edited by Michael Cole. Boston: Isabella Stewart Gardner Museum, 2014.

Pérez-Gómez, Alberto, and Louise Pelletier. *Architectural Representation and the Perspective Hinge*. Cambridge, MA: MIT Press, 2000.

Piaget, Jean, and Bärbel Inhelder. *The Child's Conception of Space*. London: Routledge & Kegan Paul, 1956.

Piero della Francesca. *De prospectiva pingendi*. New York: Broude International Editions, 1992. Written c. 1474.

Plato. *The Collected Dialogues of Plato*. Edited by Edith Hamilton and Huntington Cairns. Various translators. Princeton, NJ: Princeton University Press, 1961.

Pliny (Gaius Plinius Secundus). *Natural History: Volume 9*. Translated by H. Rackham. Cambridge, MA: Harvard University Press, 1961. Written AD 77–79.

Poe, Edgar Allan. *The Narrative of Arthur Gordon Pym of Nantucket*. New York: Harper & Brothers, 1838.

Pound, Ezra. "A Retrospect." In *Literary Essays of Ezra Pound*. New York: New Directions, 1968.

Pound, Ezra. *In a Station of the Metro*. Chicago: Poetry Foundation, 1913.

Pound, Ezra. "Vorticism." *Fortnightly Review* 96 (September 1, 1914): 461–471.

Proust, Marcel. *In Search of Lost Time*, vol. 1: *Swann's Way*. Translated by C. K. Scott Moncrieff, Terrence Kilmartin, and D. J. Enright. New York: Modern Library, 1992. First published in 1913 as *À la recherche du temps perdu*.

Ramachandran, Vilayanur S. *The Tell-Tale Brain*. New York: W. W. Norton, 2011.

Ramachandran,Vilayanur S., and Edward Hubbard. "Neurocognitive Mechanisms of Synesthesia." *Neuron* 48, no. 3 (2005): 509–520.

Retica, Aaron. "Drone-Pilot Burnout." *New York Times Sunday Magazine*, December 12, 2008.

Rich, Jack. *The Material and Methods of Sculpture*. New York: Oxford University Press, 1973.

Robson, David. "Kiki or Bouba? In Search of Language's Missing Link." *New Scientist*, July 13, 2011.

Rossi, Aldo. *A Scientific Autobiography*. Translated by Lawrence Venuti. Cambridge, MA: Opposition Books/ MIT Press, 1982. First published in 1981 as *Autobiografia scientifica*.

Rotterdam, Paul Z. *Wild Vegetation: From Art to Nature*. Munich: Hirmer Verlag, 2014.

Rowe, Colin. *As I Was Saying*, vol. 2: *Cornelliana*. Cambridge, MA: MIT Press, 1996.

Sacks, Oliver. "Neurology and the Soul." *New York Review of Books*, November 22, 1990.

Sagan, Carl. *Cosmos*. New York: Ballantine Books, 1985.

Sayce, A. H. *Introduction to the Science of Language*, vol. 1. London: K. Paul, Trench & Co., 1883.

Schelling, Friedrich Wilhelm Joseph von. *The Philosophy of Art*. Translated and commentary by Douglas W. Stott. Minneapolis: University of Minnesota, 1989. First published in 1802–1803 as *Vorlesung über die Philosophie der Kunst*.

Schjeldahl, Peter. "Improvising in Art and Life." *New York Times*, April 18, 1982.

Schmandt-Besserat, Denise. *Before Writing*, vol. 1: *From Counting to Cuneiform*. Austin: University of Texas Press, 1992.

Serra, Richard. "Verb List." In Grégoire Müller, *The New Avant-Garde: Issues for the Art of the Seventies*. New York: Praeger, 1972.

Shelley, Percy Bysshe. "A Defence of Poetry." In *The Prose Works of Percy Bysshe Shelley*. Edited by Harry Buxton Forman. London: Reeves and Turner, 1880.

Shore, Bradd. *Culture in Mind: Cognition, Culture, and the Problem of Meaning*. New York: Oxford University Press, 1998.

Simondon, Gilbert. *Du mode d'existence des objets techniques*. Paris: Aubier, 1958.

Simonyi, Károly. *A Cultural History of Physics*. Translated by David Kramer. Boca Raton, FL: AK Peters/CRC Press, 2012.

Smith, Whitney. *The Century Dictionary: An Encyclopedic Lexicon of the English Language*. New York: Century, 1914.

Soltan, Jerzy. "Working with Le Corbusier." In *Le Corbusier*. Edited by H. Allen Brooks. Princeton, NJ: Princeton University Press, 1987.

Speck, Frank G. "Reptile Lore of the Northern Indians." *Journal of American Folklore* 36, no. 141 (1923): 273–280.

Stern, William, and Clara Stern. *Psychology of Early Childhood up to the Sixth Year of Age*. Translated by Anna Barwell. Chicago: University of Chicago Press Journals, 1925.

Stevens, Wallace. *Opus Posthumous*. New York: Knopf, 1957.

Stirling, James. "Ronchamp: Le Corbusier's Chapel and the Crisis of Rationalism." *Architectural Review* (March 1956): 161.

Stout, Dietrich, Nicholas Toth, Kathy Schick, and Thierry Chaminade. "Neural Correlates of Early Stone Age Toolmaking: Technology, Language and Cognition in Human Evolution." *Philosophical Transactions of the Royal Society of London B: Biological Sciences* 363, no. 1499 (2008): 1939–1949.

Studdert-Kennedy, Michael, and Louis Goldstein. "Launching Language: The Gestural Origin of Discrete Infinity." *Studies in the Evolution of Language* 3 (2003): 235–254.

Taylor, Timothy. "Materiality." In *Handbook of Archaeological Theories*. Edited by R. Alexander Bentley, Herbert D. G. Maschner, and Christopher Chippindale. London: Rowman & Littlefield, 2008.

Thomas, Antoine Léonard. *Oeuvres de M. Thomas de l'Académie Françoise*. Amsterdam: Chez Moutard, 1773.

Tugendhat, Grete. "Letter in Response to Walter Riezler's Article 'Is the Tugendhat House Habitable?'" In Daniela Hammer-Tugendhat, Ivo Hammer, and Wolf Tegethoff, *Tugendhat House: Ludwig Mies van der Rohe*. Basel: Birkhäuser Verlag, 2015.

Tulving, Endel. "Concepts of Human Memory." In *Memory: Organization and Locus of Change*. Oxford, UK: Oxford University Press, 1991.

Uexküll, Jakob von. *Theoretical Biology*. New York: Harcourt, Brace, 1926.

Valéry, Paul. "The Idea of Art." In *Aesthetics*. Edited by Harold Osborne. Oxford, UK: Oxford University Press, 1972.

Van Gogh, Vincent. "Letter from Vincent van Gogh to Theo van Gogh, The Hague, c. 12–18 December 1882." Van Gogh's Letters. http://www.webexhibits.org/vangogh/letter/11/253.htm.

Varela, Francisco J., Evan Thompson, and Eleanor Rosch. *The Embodied Mind: Cognitive Science and Human Experience*. Cambridge, MA: MIT Press, 1991.

Virgil. *Eclogues* X. Perseus Digital Library. http://www.perseus.tufts.edu/hopper/text?doc=Perseus%3Atext%3A1999.02.0056%3Apoem%3D10.

Vygotsky, Lev. *Thought and Language.* Translated by Alex Kozulin. Cambridge, MA: MIT Press, 1986. First published in 1934 as *Mysl' I iazyk.*

Wayland, Michael. "Clay Modelers Shape the Future of Auto Design." *Chicago Tribune,* April 21, 2015.

Wegner, Daniel M. "The Mind's Best Trick: How We Experience Conscious Will." *Trends in Cognitive Sciences* 7, no. 2 (2003): 65–69.

Whitman, Walt. "A Song of Occupations." In *Leaves of Grass.* Mineola, NY: Dover, 2007. First published in 1855.

Wilde, Oscar. *The Picture of Dorian Gray.* Oxford, UK: Oxford University Press, 2008.

Williams, Lawrence E., and John A. Bargh. "Experiencing Physical Warmth Promotes Interpersonal Warmth." *Science* 322, no. 5901 (2008): 606–607.

Wilson, Frank. *The Hand.* New York: Vintage Books, 1998.

Figure Credits

Cover: Linda Pollak

Figures 1.1–1.3: Courtesy of the GM Design Archive and Special Collections

Figures 1.4–1.5: Diagrams by author

Figure 2.1: Eduardo Rivero/Shutterstock.com

Figure 3.1: Courtesy University of Pennsylvania Museum of Archaeology and Anthropology, University of Pennsylvania, Philadelphia/Denise Schmandt-Besserat

Figure 3.2: Courtesy Vorderasiatisches Museum Berlin/Denise Schmandt-Besserat

Figure 3.3: Courtesy Musée du Louvre, Département des Antiquités Orientales, Paris/Denise Schmandt-Besserat

Figure 3.4: Courtesy Cuyler Young Jr., Royal Ontario Museum, Toronto/Denise Schmandt-Besserat

Figure 4.1: William Ely Hill/Library of Congress Prints and Photographs Division

Figure 4.2: © 2018 C. Herscovici/Artists Rights Society (ARS), New York. Digital image © 2019 Museum Associates/LACMA. Licensed by Art Resource, NY

Figure 4.3: Courtesy of BBDO Belgium

Figure 4.4: Photograph by author

Figures 5.1, 5.3: Photograph © 2019 Museum of Fine Arts, Boston

Figure 5.2: © Estate of Bernd & Hilla Becher, represented by Max Becher, 2018; Courtesy Sonnabend Gallery, NY

Figure 5.4: From *The Story of Cutlery*, by J. B. Himsworth, 1953. Ernest Benn Limited, Publisher

Figures 5.5, 5.6: Drawings by Margaret Kiladjian

Figure 5.7: Photo credit: bpk Bildagentur/Staatsbibliothek zu Berlin/Dietmer Katz/Art Resource, NY

Figure 5.8: Drawing by Julie Kress

Figure 6.1: Metropolitan Museum of Art, New York

Figure 6.2: Courtesy Royal Collection Trust/Oder Teutsche Acadamie (Nuremberg 1675-9)

Figure 6.3: Château de Versailles/Wikimedia/Coyau

Figure 6.4: Photo credit: bpk Bildagentur/Gemaeldegalerie, Berlin/Jorg P. Anders/Art Resource, NY

Figure 6.5: Drawing by Julie Kress

Figure 6.6: Bayerische Staatsbibliothek, shelfmark Hbks/R 30 dx

Figure 6.7: Courtesy Sailko [CC BY 3.0 (https://creativecommons.org/licenses/by/3.0)]/Wikimedia Commons

Figure 6.8: *De prospectiva pingendi*, Piero della Francesca, *c.* 1474

Figures 6.9, 6.10: Metropolitan Museum of Art, New York

Figure 6.11: National Gallery, London/Wikimedia Commons

Figure 6.12: Graphische Sammlung Albertina. Photo credit: Erich Lessing/Art Resource, NY

Figures 6.13, 6.14: WikiArt Visual Art Encyclopedia

Figure 6.15: Courtesy of artist: Gor Chahal

Figure 7.1: Casa Buonarroti. Photo credit: Scala/Art Resource, NY

Figure 7.2: Casa Buonarroti. © Associazione Metamorfosi, Rome

Figure 7.3: Digital image courtesy of Heli Ojamaa

Figure 7.4: © F.L.C./ADAGP, Paris/Artists Rights Society (ARS), New York 2018

Figure 7.5: Digital image © Claudio Divizia/Depositphotos

Figure 7.6: Digital image © struvictory/Depositphotos

Figure 7.7: Digital image © Tate London 2019

Figure 8.1: Digital image © Tate London, © Archivio Penone. © 2018 Artists Rights Society (ARS), New York/ADAGP, Paris

Figures 8.2, 8.3, 8.5: Photos by author

Figure 8.4: Everett Collection Inc/Alamy Stock Photo

Figure 8.6: Edgar El/Wikimedia

Figure 8.7: Photo courtesy of cc-bois.com, Belgium

Figure 8.8: Photo credit: bpk Bildagentur/Hamburger Kunsthalle/Elke Walford/Art Resource, NY

Figure 8.9: Tretyakov Gallery, Moscow

Figures 8.10, 8.11: Digital image © The Museum of Modern Art/licensed by Scala/Art Resource, NY

Figure 8.12: Photo by Atelier de Sandalo. Digital image © Lempertz, Cologne

Figures 8.13–8.15, 8.18, 8.21: Photos by Atelier de Sandalo. © Rudolf Sandalo Jr., Brno City Museum

Figure 8.16: © 2018 Artists Rights Society (ARS), New York/VG Bild-Kunst, Bonn. Digital image © The Museum of Modern Art/licensed by Scala/Art Resource, NY

Figure 8.17 (top): Photo by Atelier de Sandalo. © Rudolf Sandalo Jr., Brno City Museum

Figure 8.17 (bottom): Photo by Alexandra Timpau

Figures 8.19, 8.20: Digital image © The Museum of Modern Art/licensed by Scala/Art Resource, NY

Figure 8.22: Drawing by Margaret Kiladjian

Figure 8.23: Photo: Fritz Tugendhat, 1931; courtesy of Daniela Hammer-Tugendhat

Figure 9.1: Photo by Alfred Stieglitz © Association Marcel Duchamp/ADAGP, Paris/Artists Rights Society (ARS), New York 2018

Figure 9.2: Photo: © Rheinisches Bildarchiv, Britta Schlier: rba_d033142. Museum Ludwig, Cologne, Inventar-Nr. ML 01439 © 2018 Jasper Johns/Licensed by VAGA at Artists Rights Society (ARS), NY

Figure 9.3: Adrian Phillips/Shutterstock.com

Figure 9.4: Oscar Espinosa/Shutterstock.com

Figure 9.5: ocphoto/Shutterstock.com

Figure 9.6: By K images/Shutterstock.com

Figure 9.7: Courtesy Mauro Saito, Matera

Figure 9.8: © Henri Cartier Bresson/Magnum Photos

Figure 9.9: André Kertész 1977 © 2019 Estate of André Kertész/Higher Pictures

Figure 10.1: Photograph by Seier+Seier/Wikimedia Creative Commons

Figure 10.2: Brandon Andow. Photo by author

Figure 10.3: Digital file Brandon Andow

Figures 10.4–10.7: Tae Park

Figure 10.8: © F.L.C./ADAGP, Paris/Artists Rights Society (ARS), New York 2018

Figure 11.1: Centrale Montemartin, Rome. Photo by author

Figure 11.2: Derivative photo pixelated for computer game/Aloha PK

Figure 11.3: Tida Osotsapa

Figure 11.4: Jenyea Chang

Figure 11.5: Tyler Mills

Figure 11.6: Yitan Sun

Figure 12.1: © State of New South Wales through the State Archives and Records Authority of NSW 2016

Figure 12.2: Digital image © Royal Institute of British Architects

Figure 12.3: Photo courtesy of Matter Design, 2018

Figure 12.4: Photograph Karl Dupart/Wikimedia Commons

Figure 12.5: Detroit Publishing Company/Library of Congress

Figures 12.6, 12.7: Georges Seurat/The J. Paul Getty Museum, Los Angeles

Figure 12.8: Photograph © 2019 Museum of Fine Arts, Boston

Index

Page numbers in italics indicate illustrations; those with a *t* indicate tables.

weathering effects on, 183

writing, 64–73, *66–68*, *72*

Matter Design, *Voûte de LeFebre* installation, *318*

Meaning, 62

Bruner on, 36

emotion and, 30–31

hermeneutics of, 232–233

Malafouris on, 25

of materials, 266

meaning of, 79–80

purpose and, 230–232

resistance and, 312–315, *313*, *314*

Rossi on, 229

story of, 258

Medium

definition of, 11

material as, 259–262, *262*

Medusa (Greek deity), *131*, 172–173, 218

Memory, 6–7

episodic, 49, 289

haptic, 226, 228

imagination and, 34, 44, 47, 100–102, 288

procedural, 49, 289

"pure," 34

semantic, 49, 289

storage system of, 49

visual, 42, 44–50, 83–84, 100, 288–289, 335

Memory-images, 20, 225–226, 229

Merleau-Ponty, Maurice, 21, 42, 294, 299

Metallurgy, 123, 238–239

Metaphor, 76, 80–96, 110–111, 262–263, 322

imagination and, 82–83, 87–96

neural networks of, 83–87

synesthesia and, 87

visual, 96–100, *98*, *99*

Meteorology, 295

Michelangelo Buonarroti, 161–166, *163*, *165*

Laurentian Library design, 171–172

Medici chapel at San Lorenzo, 164, *165*

modani of, 164–166, *165*, 173

Mies van der Rohe, Ludwig, 211–213

Barcelona Pavilion, 204, *206*

Illinois Institute of Technology and, 105–107, *109*

Lakeshore Drive Apartments, *110*

photograph of, *221*

Tugendhat House, 201–222, *205–221*

Mills, Tyler, *304*

Mimesis, 36, 47, 53, 60, 92, 129–130

Mind, 14

definition of, 10–11

digitization's impact on, 307–308, 331–337

enactive, 22, 103, 115, 233–235, 299, 311t, 334–336

environment and, 103

"extended," 3, 21–22, 233–235, 334

material and, 1–4, 9–10

passive, 2

Mind/body problem, 9–10, 19–25, 34, 195, 333–334

Mind's eye, 48, 78

Minimalism, 10

Mirror image, 135, 137–140

Mirror neurons, 63

Mirrors, 95, 143, 147, *148*

Mobiles, 190

Modani (architectural templates), 164–166, *165*

Models, 173–174, 272–276, *274*, *275*, 280

Moon, 5, 21, 121

Morocco, 50, 173

Mosaics, 295–299, *297*

Movement, 43, 225–229, 235–236, 299–301

bodily, 61, 155, 202, 228

in drawings, 17

extraneous, 252

smoothness of, 307, 320

Mozart, Wolfgang Amadeus, 91

Mythos/logos, 258, 267–272, *269*, *271*